Springer Series in
OPTICAL SCIENCES 117

Springer Series in
OPTICAL SCIENCES

The Springer Series in Optical Sciences, under the leadership of Editor-in-Chief *William T. Rhodes*, Georgia Institute of Technology, USA, provides an expanding selection of research monographs in all major areas of optics: lasers and quantum optics, ultrafast phenomena, optical spectroscopy techniques, optoelectronics, quantum information, information optics, applied laser technology, industrial applications, and other topics of contemporary interest.

With this broad coverage of topics, the series is of use to all research scientists and engineers who need up-to-date reference books.

The editors encourage prospective authors to correspond with them in advance of submitting a manuscript. Submission of manuscripts should be made to the Editor-in-Chief or one of the Editors. See also www.springeronline.com/series/624

Motoichi Ohtsu (Ed.)

Progress
in Nano-Electro-Optics V

Nanophotonic Fabrications,
Devices, Systems,
and Their Theoretical Bases

With 122 Figures

 Springer

Professor Dr. Motoichi Ohtsu
Department of Electronics Engineering
School of Engineering
The University of Tokyo
7-3-1 Hongo, Bunkyo-ku, Tokyo 113-8656, Japan
E-mail: ohtsu@ee.t.u-tokyo.ac.jp

ISSN 0342-4111

ISBN-10 3-540-28665-9 Springer Berlin Heidelberg New York

ISBN-13 978-3-540-28665-3 Springer Berlin Heidelberg New York

Library of Congress Cataloging-in-Publication Data

Progress in nano-electro-optics V : nanophotonic fabrications, devices, systems, and their theoretical bases /
Motoichi Ohtsu (ed.). p.cm. – (Springer series in optical sciences ; v. 117)
Includes bibliographical references and index.
ISBN 3-540-28665-9 (alk. paper)
1. Electrooptics. 2. Nanotechnology. 3. Near-field microscopy. I. Ohtsu, Motoichi. II. Series.
TA1750 .P75 2002   621.381'045–dc21   2002030321

Springer is a part of Springer Science+Business Media.

springer.com

© Springer-Verlag Berlin Heidelberg 2006
Printed in The Netherlands

Typesetting: SPI, Pondicherry, India
Cover concept by eStudio Calamar Steinen using a background picture from The Optics Project. Courtesy of John T. Foley, Professor, Department of Physics and Astronomy, Mississippi State University, USA.
Cover production: *design & production* GmbH, Heidelberg

Printed on acid-free paper   SPIN: 11540205   57/3100/SPI 5 4 3 2 1 0

Preface to *Progress in Nanoelectro-Optics*

Recent advances in electro-optical systems demand drastic increases in the degree of integration of photonic and electronic devices for large-capacity and ultrahigh-speed signal transmission and information processing. Device size has to be scaled down to nanometric dimensions to meet this requirement, which will become even more strict in the future. In the case of photonic devices, this requirement cannot be met only by decreasing the sizes of materials. It is indispensable to decrease the size of the electromagnetic field used as a carrier for signal transmission. Such a decrease in the size of the electromagnetic field beyond the diffraction limit of the propagating field can be realized in optical near fields.

Near-field optics has progressed rapidly in elucidating the science and technology of such fields. Exploiting an essential feature of optical near fields, i.e., the resonant interaction between electromagnetic fields and matter in nanometric regions, important applications and new directions such as studies in spatially resolved spectroscopy, nanofabrication, nanophotonic devices, ultrahigh-density optical memory, and atom manipulation have been realized and significant progress has been reported. Since nanotechnology for fabricating nanometric materials has progressed simultaneously, combining the products of these studies can open new fields to meet the above-described requirements of future technologies.

This unique monograph series entitled "Progress in Nanoelectro-Optics" is being introduced to review the results of advanced studies in the field of electro-optics at nanometric scales and covers the most recent topics of theoretical and experimental interest on relevant fields of study (e.g., classical and quantum optics, organic and inorganic material science and technology, surface science, spectroscopy, atom manipulation, photonics, and electronics). Each chapter is written by leading scientists in the relevant field. Thus, high-quality scientific and technical information is provided to scientists, engineers, and students who are and will be engaged in nanoelectro-optics and nanophotonics research.

I am gratefull to the members of the editorial advisory board for their valuable suggestions and comments in organizing this monograph series. I wish to express my special thanks to Dr. T. Asakura, editor of the Springer Series in Optical Sciences and Professor Emeritus, Hokkaido University for recommending me to publish this monograph series. Finally, I extend my acknowledgement to Dr. Claus Ascheron of Springer-Verlag, for his guidance and suggestions, and to Dr. H. Ito, associate editor, for his assistance throughout the preparation of this monograph series.

Yokohama, October 2002   *Motoichi Ohtsu*

Preface to Volume V

This volume contains four review articles focusing on nanophotonics. Nanophotonics has been proposed by M. Ohtsu in 1993, which is a novel technology that utilizes local electromagnetic interactions between a few nanometric objects and an optical near field. Since an optical near field is free from the diffraction of light due to its size-dependent localization and size-dependent resonance features, nanophotonics enables fabrication and operation of nanometric devices. Further, ultrahigh-capacity information processing systems are possible by integrating these devices. However, it should be noted that nanophotonics is not only to realize nanometer-sized optical technology (quantitative innovation). It can realize novel functions and phenomena, which are not possible as long as propagating lights are used (qualitative innovation). Producing the qualitative innovation is the significance of nanophotonics, i.e., prominent advantages over conventional photonics. Due to the qualitative innovation, nanophotonics is expected to shift the paradigm of optical industry and market.

Nanophotonics is closely related to quantum optics, atom optics, nanostructure fabrication technology, information processing system, and so on. And for this relationship, nanophotonics exhibits rapid progress in these years.

To establish theoretical bases of the rapidly progressing nanophotonics, the first article reviews theories for operation principles of characteristic nanophotonic functional devices, in which optical near fields play roles of information carriers and control signals. Dynamics of these devices are also studied as well as formulated characteristic interaction between nanometric objects and an optical near field.

The second article aims at describing the optical near-field phenomena and their applications to fabricate nanophotonic devices. To realize nanometer-scale controllability in size and position, the feasibility of nanometer-scale chemical vapor deposition is demonstrated using optical near-field techniques. Further, fabrication and operation of nanometer-scale waveguides are demonstrated, which are used as conversion devices of the nanophotonic integrated circuits.

The third article summarizes unique properties of optical near fields, i.e., optically forbidden energy transfer between quantum dots, anti-parallel dipole coupling of quantum dots, and nonadiabatic photochemical interactions. Their applications to nanophotonic devices and nanofabrication are demonstrated.

The last article concerns nanophotonic information and communications systems. They can overcome the integration-density limit with ultralow-power operation as well as unique functionalities, which are only achievable using optical near-field interactions. Two architectural approaches to these systems are discussed. One is a memory-based architecture which is based on table lookup using optical near-field interactions between quantum dots. Another is one focusing on hierarchy. As an example, a hierarchical memory system is presented.

As was the case of volumes I–IV, this volume is published with the support of an associate editor and members of editorial advisory board. They are:

Associate editor: Ito, H. (Tokyo Inst. Tech., Japan)
Editorial advisory board: Barbara, P.F. (Univ. of Texas, USA)
                         Bernt, R. (Univ. of Kiel, Germany)
                         Courjon, D. (Univ. de Franche-Comte, France)
                         Hori, H. (Univ. of Yamanashi, Japan)
                         Kawata, S. (Osaka Univ., Japan)
                         Pohl, D. (Univ. of Basel, Switzerland)
                         Tsukada, M. (Univ. of Tokyo, Japan)
                         Zhu, X. (Peking Univ., China)

I hope that this volume will be a valuable resource for the readers and future specialists.

Tokyo, June 2005   *Motoichi Ohtsu*

Contents

Architectural Approach to Nanophotonics
for Information and Communication Systems

M. Naruse, T. Kawazoe, T. Yatsui, S. Sangu, K. Kobayashi, M. Ohtsu . 163

List of Contributors

Tadashi Kawazoe
Solution-Oriented Research
for Science and Technology
Japan Science and Technology
Agency
687-1 Tsuruma, Machida
Tokyo 194-0004, Japan
kawazoe@ohtsu.jst.go.jp

Kiyoshi Kobayashi
Department of Physics
Tokyo Institute of Technology
2-12-1-H79 O-okayama, Meguro-ku
Tokyo 152-8551, Japan
kkoba@phys.titech.ac.jp

Makoto Naruse
New Generation Network
Research Center
National Institute of Information
and Communications Technology
4-2-1 Nukui-kita, Koganei
Tokyo 184-8795, Japan
naruse@nict.go.jp

Arup Neogi
Department of Physics
University of North
Texas
Denton, TX 76203, USA
arup@unt.edu

Motoichi Ohtsu
School of Engineering
The University of Tokyo
7-3-1 Hongo, Bunkyo-ku
Tokyo 113-8656, Japan
ohtsu@ee.t.u-tokyo.ac.jp

Suguru Sangu
Advanced Technology R&D Center
Ricoh Co. Ltd.
16-1 Shinei-cho, Tsuzuki-ku
Yokohama 224-0035, Japan
suguru.sangu@nts.ricoh.co.jp

Akira Shojiguchi
Department of Physics
Nara Women's University
Kitauoyanishi-machi
Nara 630-8506, Japan
shojiguchi@ki-rin.phys.
nara-wu.ac.jp

Takashi Yatsui
Solution-Oriented Research
for Science and Technology
Japan Science and Technology
Agency
687-1 Tsuruma, Machida
Tokyo 194-0004, Japan
yatsui@ohtsu.jst.go.jp

Gyu-Chul Yi
Department of Materials Science
and Engineering
Pohang University of Science
and Technology
San 31 Hyoja-dong, Pohang
Gyeongbuk 790-784, Korea
gcyi@postech.ac.kr

Theory and Principles of Operation
of Nanophotonic Functional Devices

S. Sangu, K. Kobayashi, A. Shojiguchi, T. Kawazoe, and M. Ohtsu

1 Introduction

1.1 Nanophotonics for Functional Devices

In response of the need for increased and faster information processing in the near future, miniaturization of optical devices has progressed [1] to the point that it has now almost reached the critical limit determined by the diffraction of conventional propagating light [2,3]. Since 1990s, researchers have anticipated that optical near-field devices may be one of the first important technologies to overcome this limit; many studies have been performed in various fields such as fundamental physics in nanometric space, optical near-field microscopy and spectroscopy, optical measurement, bioimaging, nanofabrication, and nanophotonic device architecture [4]. An optical near field is the characteristic localized electromagnetic field around a nanometric object, and its decay length, which is smaller than the wavelength of incident light, depends on the size of the object. This size dependence means that optical near fields cannot be separated from matter excitation; in nanometrics pace, the incident electromagnetic field is modified by matter excitation in an object, and the modified field also affects the object itself and another neighboring one before releasing the energy as far-field photons. This nanometric light–matter interaction must describe as a self-consistent field. The goal is to create nanometric functional devices that are free from light diffraction limits, in which such optical near fields act as information carrier and control signals. These devices are termed *nanophotonic devices*. The localization feature of nanophotonic devices seems to resemble electronic devices in which an electric charge always stays within the device, but in a nanophotonic device, the localized field is able to leave an object and release photons in the far field via optical near-field interaction among several nanometric objects [5]. An important component of nanophotonic devices and nanophotonic device operations is dealing with light–matter interaction with a nanometric system, as well as dissipation of matter excitation energy toward the outer field. Since the signal is eventually detected as far-field light, nanometric light–matter interaction also needs to control the dissipation process. Hence, the inherent operation of a nanophotonic device in nanophotonics differs from conventional optical and electronic devices.

The advantages of nanophotonic devices include not only miniaturization but also possibilities for novel principles of functional operations that are inherent to nanophotonics. As mentioned above, the physics of nanophotonic devices includes typical matter excited states due to optical near-field interaction, coupling between near- and far-field light, and coupling between matter excitation and phonons. Many of these characteristics have not been considered in conventional optics; the structures within nanophotonic devices may differ from those of conventional devices, since basic principles utilized differ and nanophotonic devices can accomplish functions that have not been possible to date. It is important to consider how these devices should be designed, and to learn how nanophotonic devices can coexist with other devices.

In this chapter, our discussion focuses on how to use the features inherent to nanophotonics in functional device operations, what is possible, and how we can realize the possibilities. Section 1.2 explains some characteristic features of nanophotonics and provides a basic outline of nanophotonic devices.

1.2 Inherent Features to Nanophotonics

In general, the following features are indispensable to a functional device: preparation of appropriate input states, propagation of a signal, and control of the signal. Nanophotonics has characteristic features for all of these, none of which are observed in far-field light. This section explains features inherent to nanophotonics: a locally excited state that cannot be created using far-field light, unidirectional energy transfer, and a dependence on excitation number in which coupling between discrete energy levels and the optical near field plays an important role. These are all key features for nanophotonic device operations.

Locally Excited States

First, we will explain the difference between matter excitation of nanometric objects using far- and near-field light. Figure 1a illustrates nanometric objects being irradiated by far-field light. Since objects located in an area much smaller than the wavelength of the light are simultaneously excited by a uniform field, it is neither possible to examine the state of matter excitation in each object independently nor does the detected far-field light provide any information about the state of excitement in each object. Optical near-field excitation can be accomplished by setting an optical near-field probe, such as a nanometric metallic aperture, an optical fiber probe, and a single molecule. This allows selective irradiation of individual object, and a locally excited

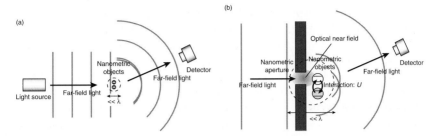

Fig. 1. Schematic illustration of nanometric matter excitation by using **(a)** far-field light and **(b)** near-field light

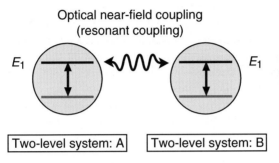

Fig. 2. Identical two two-level systems which are coupled via optical near-field interaction

state can be created because of the localized light around the probe. The excited object creates a secondary electromagnetic field that affects neighboring fields via the optical near field, and consistently determining excited states in the system using optical near-field interaction. Figure 1b shows schematics of optical near-field excitation. This asymmetric excitation also influences far-field light which can be detected as an information signal.

The following describes these excited states as algebraic expressions. For simplicity, the discussion is restricted to the coupling of two two-level systems of excitons (see Fig. 2). For far-field excitation, nanometric objects are uniformly excited; a one-exciton state, in which an exciton exists in the system, can thus be written as

$$|1\rangle_s = (|e\rangle_A |g\rangle_B + |g\rangle_A |e\rangle_B)/\sqrt{2} \,, \tag{1}$$

where $|e\rangle_i$ and $|g\rangle_i$ represent exciton state and crystal ground state, respectively. Subscripts, $i = A$ and B, label two nanometric objects, and the meaning of subscript s will be explained later. Equation (1) means that an exciton in an isolated system cannot be distinguished because the exciton exists in both object A and object B with equivalent probabilities.

On the other hand, as mentioned above, an optical near field allows an exciton to be created in an individual object. The exciton prepared in this system leaves and returns between the two two-level systems for a period depending on the strength of optical near-field coupling, referred to as near-field optical nutation [6,7]. However, if the pumping time is much shorter than the period of near-field optical nutation, locally excited states can be created in this system. Such locally excited states with an exciton in the system can be expressed by a linear combination of coupled states that extends between two objects, as follows:

$$|e\rangle_A |g\rangle_B = (|1\rangle_s + |1\rangle_a)/\sqrt{2}\,, \tag{2}$$

$$|g\rangle_A |e\rangle_B = (|1\rangle_s - |1\rangle_a)/\sqrt{2}\,. \tag{3}$$

The right-hand terms in (2) and (3) described states coupled via an optical near field, where the subscripts s and a refer to symmetric and anti-symmetric states, respectively. It is clear that in the optical near-field excitation, there are two coupled states while far-field light excites only the symmetric state. Note that we did not show the anti-symmetric state in (1), since the state is optically inactive for far-field light. This can be verified using the following relation: $_a\langle 1|\widehat{H}_{int}|g\rangle = 0$, where $|g\rangle = |g\rangle_A |g\rangle_B$ and the interaction Hamiltonian refers to (5). Locally excited states are quite important for functional operations in our proposed nanophotonic devices, which are discussed in Sects. 3 and 4.

Unidirectional Energy Transfer

For functional device operations to manipulate information carriers, an excitation or carrier must transfer unidirectionally from the input to the output terminals. In conventional optical devices, a unidirectional energy transfer can be accomplished by using an optical isolator, which generally uses polarization to block reflected light. Unless polarization is used, the size of optical devices is restricted by light wavelength. In electronic devices, a unidirectional signal transfer is easily attained since electrons flow along an electrical potential. However, as electronic devices become smaller and quantum mechanical effects arise, electrical signals are affected by noise because of universal quantum fluctuations. In a nanophotonic device, signal isolation using light wave characteristics is impossible because of the light diffraction limit, and a signal carrier is composed of electrically neutral quasi-particles of electrons and holes. Thus, a static electrical potential cannot be used to drive them. However, unidirectional exciton energy transfer can be effectively realized using a relaxation process among quantum discrete energy levels [8]. Figure 3 is a schematic image of energy transfer via an optical near field in a system that consists of two nanometric objects with two- and three-energy levels. As mentioned in "Locally Excited States," optical near-field coupling causes a coherently coupled excited state between the E_1-level in the two-level system and the E_2-level in the three-level system, which strengthens when both energies are equal. If excitation can be dropped into the lower E_1-level in the three-level system before the radiative lifetime of E_1-level in the two-level system ($\sim$1 ns), excitation is confined to the energy level due to off-resonance,

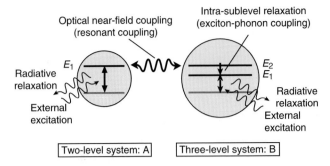

Fig. 3. Energy transfer between two-level and three-level systems. E_2-level in the three-level system is dipole inactive, and thus, the unidirectional energy transfer is achieved only by mediating an optical near field

and irreversibility in the nanometric system is guaranteed except for radiation from the energy level. Section 2 provides a detailed discussion about optical near-field coupling and energy transfer dynamics. In the three-level system, the E_2-level is generally dipole inactive for far-field light, and thus, unidirectional energy transfer can achieved by mediating the optical near field.

Since external far- or near-field light can cause excitations in dipole active levels: the E_1-levels in the two-level and the three-level systems, energy transfer in this system is controllable. A simple switching operation can be constructed using the state-filling nature of excitons excited by the external field. In Sect. 3, a nanophotonic switch that uses energy transfer and state-filling is proposed, and the dynamics of excitation are evaluated both analytically and numerically.

Dependence of Excitation Number

Although symmetric and anti-symmetric states in (1)–(3) describe one-exciton states, a quite interesting feature is evident in the two-exciton state in the system shown in Fig. 2. The two-exciton state, in which two excitons completely occupy both two-level systems, is algebraically written as

$$|2\rangle_p = |e\rangle_A |e\rangle_B , \tag{4}$$

where number 2 in the left-hand side refers to the two-exciton state. It is valuable to investigate energies for all base states, $|1\rangle_s$, $|1\rangle_a$, and $|2\rangle_p$. The Hamiltonian for the two-level systems coupled via an optical near-field interaction is given by

$$\widehat{H} = \widehat{H}_0 + \widehat{H}_{int} , \tag{5}$$

$$\widehat{H}_0 = \hbar\Omega\widehat{A}^\dagger\widehat{A} + \hbar\Omega\widehat{B}^\dagger\widehat{B} , \tag{6}$$

$$\widehat{H}_{int} = \hbar U(\widehat{A}^\dagger\widehat{B} + \widehat{A}\widehat{B}^\dagger) , \tag{7}$$

where $\widehat{H}_0$ and $\widehat{H}_{int}$ represent the unperturbed and interaction Hamiltonian, respectively. $(\widehat{A}^\dagger, \widehat{A})$ and $(\widehat{B}^\dagger, \widehat{B})$ are the fermionic creation and annihilation

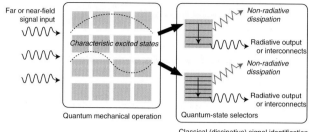

Fig. 4. Conceptual structure of nanophotonic devices, which consists of a quantum mechanical part and a classical dissipative part. Quantum mechanical part builds up characteristic excited states and classical dissipative part identifies certain states and connects to outer detection systems

operators in the two-level systems A and B, respectively. Since the excitations are assumed to be fermionic excitons, and the optical near-field coupling U is considered a completely coherent process; this is explained in detail in Sect. 2. Energies for states are given as follows:

$$_s\langle 1|\widehat{H}|1\rangle_s = \hbar(\Omega + U)\,, \tag{8}$$

$$_a\langle 1|\widehat{H}|1\rangle_a = \hbar(\Omega - U)\,, \tag{9}$$

$$_p\langle 2|\widehat{H}|2\rangle_p = 2\hbar\Omega\,. \tag{10}$$

Equations (8) and (9) indicate that the energies of the coupled states, $|1\rangle_s$ and $|1\rangle_a$, depend on the strength of optical near-field coupling, U, and differences in the energy from the two-level system have opposite contributions in each state. In the two-exciton states in (10), energy apparently degenerates because both systems are completely filled. These properties are useful for selective energy transfer in nanophotonic devices; sequential logic operations can be realized by using the excitation number dependence in this system.

Figure 4 schematically illustrates how the above selectivity represents a concept that is fundamental to nanophotonic devices. In the device, quantum mechanical and classical parts coexist; some characteristic excited states are created in the quantum mechanical part, and in order to connect a signal to an outer detection system, these states must then be selectively extracted from quantum mechanical part to classical dissipative one. This process is key to driving the nanophotonic device. Functional operations based on such conceptual structures are discussed in Sect. 4.

2 Optical Near-Field Coupling

In this section, we give a full account of energy transfer between locally excited states via an optical near field. From our theoretical treatment of optical near-field coupling, the readers will understand why dipole-inactive energy transfer for far-field light changes allowed transition in the case of the optical near field. Concrete numerical results of coupling strength in a CuCl quantum-dot system are also provided, where the coupling strength determines operation speed of nanophotonic devices discussed in Sects. 3 and 4.

2.1 Theoretical Descriptions of an Optical Near Field

There are two ways to describe light–matter interaction theoretically; one is to use the minimal coupling Hamiltonian $\boldsymbol{p} \cdot \boldsymbol{A}$ [9], $\boldsymbol{p}$ being the electronic momentum and $\boldsymbol{A}$ the vector potential, and the other is to use the multipolar QED Hamiltonian [10, 11] in the dipole approximation, $\boldsymbol{\mu} \cdot \boldsymbol{D}$, where $\boldsymbol{\mu}$ and $\boldsymbol{D}$ represent the electric dipole moment and electric displacement field, respectively. The two descriptions of light–matter interaction are connected by Power-Zienau–Woolley transformation [12], which is a unitary transformation of the Coulomb-gauge Hamiltonian. Here, the multipolar QED Hamiltonian is used because there are several advantages in the multipolar QED; first of all, it does not contain any explicit intermolecular or interquantum-dot Coulomb interactions in the interaction Hamiltonian and entire contribution to the fully retarded result is from the exchange of transverse photons, while in the minimal coupling, the intermolecular interactions arise both from the exchange of transverse photons, which include static components, and from the instantaneous intermolecular electrostatic interactions [13]. Second, it clarifies physical interpretation of the dipole inactive transition via the optical near field as we will discuss later.

Since our purpose of discussions is to propose and investigate nanophotonic solid devices, in the following, nanometric objects are assumed as quantum dots with discrete energy levels. In order to explain an extremely important feature in nanophotonics, internal electronic structures in a quantum dot are regarded as a collection of local dipoles, which is convenient to express the interactions between nanometric objects and an optical near field spatially distributed in nanometric space. We can also depict a dipole in one-body problem by using an effective mass approximation. Such theoretical approach has already been published [14] where the enhancement of electric quadrupole coupling was pointed out by assuming steep variation of electric field due to the optical near field. This phenomenon is equivalent to our result of the dipole-inactive transition, but in our theoretical formulation, in which field variation is caused by the coupling between the local dipoles in the neighboring quantum-dot pair, it is easy to obtain a physical interpretation.

In Sect. 2.2 interaction Hamiltonian is provided in the second-quantized form in terms of electron basis functions satisfying the quantum-dot boundary conditions, as well as transition dipole moments of excitons, and then, optical near-field coupling is derived on the basis of the projection operator method which is explained in Sect. 2.3.

2.2 Excitation and Transition in a Quantum Dot

Interaction Hamiltonian

According to the dipole coupling in the multipolar QED Hamiltonian, the interaction between photons and nanometric materials can be written as [11]

$$\widehat{H}_{\text{int}} = -\int \psi^{\dagger}(\boldsymbol{r})\boldsymbol{\mu}(\boldsymbol{r})\psi(\boldsymbol{r}) \cdot \widehat{\boldsymbol{D}}(\boldsymbol{r}) \mathrm{d}\boldsymbol{r} \ , \tag{11}$$

where $\psi^{\dagger}(\boldsymbol{r})$ and $\psi(\boldsymbol{r})$ denote field operators for electron creation and annihilation, respectively, and the dipole moment and the second-quantized electric displacement vector at position $\boldsymbol{r}$ are expressed as $\boldsymbol{\mu}(\boldsymbol{r})$ and $\widehat{\boldsymbol{D}}(\boldsymbol{r})$, respectively. In a quantum dot, the electron field operators should be expanded in terms of basis functions $\phi_{\nu\boldsymbol{n}}(\boldsymbol{r})$ that satisfy the electron boundary conditions in a quantum dot, that is analogy to those in bulk materials where the Bloch functions satisfying periodic boundary condition are used. The field operators are given by

$$\psi(\boldsymbol{r}) = \sum_{\nu=\mathrm{c,v}} \sum_{\boldsymbol{n}} \hat{c}_{\nu\boldsymbol{n}} \phi_{\nu\boldsymbol{n}}(\boldsymbol{r}) \ , \tag{12}$$

$$\psi^{\dagger}(\boldsymbol{r}) = \sum_{\nu=\mathrm{c,v}} \sum_{\boldsymbol{n}} \hat{c}^{\dagger}_{\nu\boldsymbol{n}} \phi^{*}_{\nu\boldsymbol{n}}(\boldsymbol{r}) \ , \tag{13}$$

where $\hat{c}^{\dagger}_{\nu\boldsymbol{n}}$ and $\hat{c}_{\nu\boldsymbol{n}}$ represent the creation and annihilation operators for the electrons specified by $(\nu, \boldsymbol{n})$, respectively, and the indices $\nu = \mathrm{c, v}$ denote the conduction and valence bands. The discrete energy levels in the quantum dot are labeled $\boldsymbol{n}$. The basis functions satisfy the following completeness condition, as well as orthonormalization:

$$\sum_{\nu=\mathrm{c,v}} \sum_{\boldsymbol{n}} \phi^{*}_{\nu\boldsymbol{n}}(\boldsymbol{r})\phi_{\nu\boldsymbol{n}}(\boldsymbol{r}') = \delta(\boldsymbol{r} - \boldsymbol{r}') \ . \tag{14}$$

Simultaneously, we express the electric displacement vector $\widehat{\boldsymbol{D}}(\boldsymbol{r})$ using exciton–polariton creation and annihilation operators $(\hat{\xi}^{\dagger}_{\boldsymbol{k}}, \hat{\xi}_{\boldsymbol{k}})$, where branch suffix of the exciton–polariton is suppressed by taking only an upper branch. We consider exciton–polaritons because a nanometric system in a near-field optical environment is always surrounded by macroscopic materials, such as the substrate, matrix, fiber probe, and so on. Previously [15,16], we proposed an effective interaction for such a nanometric system mediated by exciton–polaritons that exists in mixed states between photons and macroscopic material excitations instead of free photons. We showed that such a treatment provides a good description of the characteristics of an optical near field [17]. Using this, the electric displacement vector $\widehat{\boldsymbol{D}}(\boldsymbol{r})$ in (11) can be written as [18]

$$\widehat{\boldsymbol{D}}(\boldsymbol{r}) = \mathrm{i}\sqrt{\frac{2\pi}{V}} \sum_{\boldsymbol{k}} \sum_{\lambda=1}^{2} \boldsymbol{e}_{\lambda}(\boldsymbol{k})f(k)(\hat{\xi}_{\boldsymbol{k}}\mathrm{e}^{\mathrm{i}\boldsymbol{k}\cdot\boldsymbol{r}} - \hat{\xi}^{\dagger}_{\boldsymbol{k}}\mathrm{e}^{-\mathrm{i}\boldsymbol{k}\cdot\boldsymbol{r}}) \tag{15}$$

with

$$f(k) = \frac{\hbar c k}{\sqrt{E(k)}} \sqrt{\frac{E^2(k) - E_{\mathrm{m}}^2}{2E^2(k) - E_{\mathrm{m}}^2 - \hbar^2 c^2 k^2}} \, , \qquad (16)$$

where $\hbar$, V, $\boldsymbol{e}_\lambda(\boldsymbol{k})$, and $\boldsymbol{k}$ are the Dirac constant, the quantization volume, the unit polarization vector, and the wavevector of the exciton–polaritons, respectively. Here we assume $\boldsymbol{e}_\lambda(\boldsymbol{k})$ as real. The speed of light in a vacuum is c, and the exciton–polariton energy with a wavevector $\boldsymbol{k}$ and the macroscopic material excitation energy are $E(\boldsymbol{k})$ and E_{m}, respectively. Substituting (12) and (15) into (11) gives the interaction Hamiltonian in the second-quantized representation as

$$\widehat{H}_{\mathrm{int}} = \sum_{\nu n \nu' n' k \lambda} (\hat{c}_{\nu n}^\dagger \hat{c}_{\nu' n'} \hat{\xi}_k g_{\nu n \nu' n' k \lambda} - \hat{c}_{\nu n}^\dagger \hat{c}_{\nu' n'} \hat{\xi}_k^\dagger g_{\nu n \nu' n' - k \lambda}) \qquad (17)$$

with

$$g_{\nu n \nu' n' k \lambda} = -\mathrm{i}\sqrt{\frac{2\pi}{V}} f(k) \int \phi_{\nu n}^*(\boldsymbol{r})(\boldsymbol{\mu}(\boldsymbol{r}) \cdot \boldsymbol{e}_\lambda(\boldsymbol{k})) e^{\mathrm{i} \boldsymbol{k} \cdot \boldsymbol{r}} \phi_{\nu' n'}(\boldsymbol{r}) \mathrm{d}\boldsymbol{r} \, . \qquad (18)$$

Transition Matrix Element for Exciton States

In order to describe the creation and annihilation of excitons in a quantum dot, it is convenient to use the Wannier representation in which electrons are localized in an atomic site $\boldsymbol{R}$. Then, the electron field operators can be expanded using the Wannier functions $w_{\nu \boldsymbol{R}}(\boldsymbol{r})$ instead of $\phi_{\nu n}(\boldsymbol{r})$

$$\psi(\boldsymbol{r}) = \sum_{\nu = \mathrm{c,v}} \sum_{\boldsymbol{R}} \hat{c}_{\nu \boldsymbol{R}} w_{\nu \boldsymbol{R}}(\boldsymbol{r}) \, , \quad \psi^\dagger(\boldsymbol{r}) = \sum_{\nu = \mathrm{c,v}} \sum_{\boldsymbol{R}} \hat{c}_{\nu \boldsymbol{R}}^\dagger w_{\nu \boldsymbol{R}}^*(\boldsymbol{r}) \, , \qquad (19)$$

where $c_{\nu \boldsymbol{R}}^\dagger$ and $c_{\nu \boldsymbol{R}}$ denote the creation and annihilation operators of electrons at site $\boldsymbol{R}$ in the energy band ν. These operators in the Wannier representation are written in terms of $\hat{c}_{\nu n}$ and $\hat{c}_{\nu n}^\dagger$ in (12) as follows:

$$\hat{c}_{\nu \boldsymbol{R}} = \sum_{\nu' = \mathrm{c,v}} \sum_{n} \hat{c}_{\nu' n} \int w_{\nu \boldsymbol{R}}^*(\boldsymbol{r}) \phi_{\nu' n}(\boldsymbol{r}) \, \mathrm{d}\boldsymbol{r} \, , \qquad (20)$$

$$\hat{c}_{\nu \boldsymbol{R}}^\dagger = \sum_{\nu' = \mathrm{c,v}} \sum_{n} \hat{c}_{\nu' n}^\dagger \int w_{\nu \boldsymbol{R}}(\boldsymbol{r}) \phi_{\nu' n}^*(\boldsymbol{r}) \, \mathrm{d}\boldsymbol{r} \, . \qquad (21)$$

When we assume excitons in the weak-confinement regime, i.e., an exciton Bohr radius to be smaller than the quantum-dot size, the exciton states in a quantum dot specified by the quantum number $\boldsymbol{m}$ and μ can be described by superposition of the excitons in the Wannier representation as [19]

$$\begin{aligned} |\Phi_{\boldsymbol{m} \mu}\rangle &= \sum_{\boldsymbol{R}, \boldsymbol{R}'} F_{\boldsymbol{m}}(\boldsymbol{R}_{\mathrm{CM}}) \varphi_\mu(\boldsymbol{\beta}) \hat{c}_{\mathrm{c}\boldsymbol{R}'}^\dagger \hat{c}_{\mathrm{v}\boldsymbol{R}} |\Phi_{\mathrm{g}}\rangle \\ &= \sum_{\boldsymbol{R}, \boldsymbol{R}'} F_{\boldsymbol{m}}(\boldsymbol{R}_{\mathrm{CM}}) \varphi_\mu(\boldsymbol{\beta}) \sum_{n, n'} h_{\boldsymbol{R} n \boldsymbol{R}' n'} \hat{c}_{\mathrm{c}n}^\dagger \hat{c}_{\mathrm{v}n'} |\Phi_{\mathrm{g}}\rangle \, , \end{aligned} \qquad (22)$$

where $F_m(\boldsymbol{R}_{\mathrm{CM}})$ and $\varphi_\mu(\boldsymbol{\beta})$ denote the envelope functions for the center of mass and relative motions of the excitons, respectively. These are $\boldsymbol{R}_{\mathrm{CM}} = (m_e\boldsymbol{R}' + m_h\boldsymbol{R})/(m_e + m_h)$ and $\boldsymbol{\beta} = \boldsymbol{R}' - \boldsymbol{R}$, where m_e and m_h are the effective masses of the electrons and holes. The overlap integrals $h_{\boldsymbol{R}n\boldsymbol{R}'n'}$ are defined as

$$h_{\boldsymbol{R}n\boldsymbol{R}'n'} = \iint w_{v\boldsymbol{R}}^*(\boldsymbol{r}_2)w_{c\boldsymbol{R}'}(\boldsymbol{r}_1)\phi_{cn}^*(\boldsymbol{r}_1)\phi_{vn'}(\boldsymbol{r}_2)\mathrm{d}\boldsymbol{r}_1\,\mathrm{d}\boldsymbol{r}_2 \ . \tag{23}$$

The sum of ν' in (20) is determined automatically as $\hat{c}_{cn}^\dagger$ and $\hat{c}_{vn'}$ because the valence band is fully occupied in the initial ground state $|\Phi_g\rangle$. Using (17) and (22), the transition matrix element from the exciton state to the ground state is obtained as

$$\langle\Phi_g|\widehat{H}_{\mathrm{int}}|\Phi_{m\mu}\rangle = \sum_{n_1,n_2}\sum_{\boldsymbol{R},\boldsymbol{R}'} F_m(\boldsymbol{R}_{\mathrm{CM}})\varphi_\mu(\boldsymbol{\beta})$$

$$\times \sum_{\boldsymbol{k}}\sum_{\lambda=1}^2(\hat{\xi}_{\boldsymbol{k}}g_{vn_1cn_2\boldsymbol{k}\lambda} - \hat{\xi}_{\boldsymbol{k}}^\dagger g_{vn_1cn_2-\boldsymbol{k}\lambda})h_{\boldsymbol{R}n_2\boldsymbol{R}'n_1} \ , \tag{24}$$

where we use the following relation:

$$\langle\Phi_g|\hat{c}_{vn_1}^\dagger\hat{c}_{cn_2}\hat{c}_{cn_3}^\dagger\hat{c}_{vn_4}|\Phi_g\rangle = \delta_{n_1,n_4}\delta_{n_2,n_3} \ . \tag{25}$$

In addition, with the help of the completeness and orthonormalization of $\phi_{\nu n}(\boldsymbol{r})$ [see (14)], we can simplify the product of g and h as

$$\sum_{n_1,n_2} g_{vn_1cn_2\boldsymbol{k}\lambda}h_{\boldsymbol{R}n_2\boldsymbol{R}'n_1} = -i\sqrt{\frac{2\pi}{V}}f(k)\int w_{v\boldsymbol{R}}^*(\boldsymbol{r})\boldsymbol{\mu}(\boldsymbol{r})w_{c\boldsymbol{R}'}(\boldsymbol{r})\cdot\boldsymbol{e}_\lambda(\boldsymbol{k})e^{i\boldsymbol{k}\cdot\boldsymbol{r}}\mathrm{d}\boldsymbol{r}$$

$$\approx -i\sqrt{\frac{2\pi}{V}}f(k)(\boldsymbol{\mu}_{cv}\cdot\boldsymbol{e}_\lambda(\boldsymbol{k}))e^{i\boldsymbol{k}\cdot\boldsymbol{R}}\delta_{\boldsymbol{R},\boldsymbol{R}'} \ , \tag{26}$$

where the transformation of the spatial integral in the first line of (26) into the sum of the unit cells and the spatial localization of the Wannier functions provides $\delta_{\boldsymbol{R},\boldsymbol{R}'}$ in the second line. The transition dipole moment for each unit cell is defined as

$$\boldsymbol{\mu}_{cv} = \int_{\mathrm{UC}} w_{v\boldsymbol{R}}^*(\boldsymbol{r})\boldsymbol{\mu}(\boldsymbol{r})w_{c\boldsymbol{R}}(\boldsymbol{r})\mathrm{d}\boldsymbol{r} \ . \tag{27}$$

We assume that the transition dipole moment is the same as that of the bulk material, independent of the site $\boldsymbol{R}$, and that the electric displacement vector is uniform at each site. Finally, (24) is reduced to

$$\langle\Phi_g|\widehat{H}_{\mathrm{int}}|\Phi_{m\mu}\rangle = -i\sqrt{\frac{2\pi}{V}}\sum_{\boldsymbol{R}}\sum_{\boldsymbol{k}}\sum_{\lambda=1}^2 f(k)(\boldsymbol{\mu}_{cv}\cdot\boldsymbol{e}_\lambda(\boldsymbol{k}))F_m(\boldsymbol{R})\varphi_\mu(0)$$

$$\times\left(\hat{\xi}_{\boldsymbol{k}}e^{i\boldsymbol{k}\cdot\boldsymbol{R}} - \hat{\xi}_{\boldsymbol{k}}^\dagger e^{-i\boldsymbol{k}\cdot\boldsymbol{R}}\right) \ . \tag{28}$$

Here, we note that the exciton–polariton field expanded by the plane wave with the wavevector $\boldsymbol{k}$ depends on the site $\boldsymbol{R}$ in the quantum dot because we do not apply the long wave approximation that is usually used for far-field light.

2.3 Optical Near-Field Coupling Between Quantum Dots

Formulation

Until now, we have derived the transition matrix element from the exciton state to the ground state in a quantum dot. Considering operations of nanophotonic devices, signal carrier corresponds to the energy transfer between nanometric objects, or quantum dots, which are electronically separated, and the speed of the energy transfer is determined by the coupling strength of an optical near field. In this stage, we derive the coupling strength

$$\hbar U = \langle \Psi_f | \widehat{H}_{int} | \Psi_i \rangle , \tag{29}$$

where $|\Psi_i\rangle$ and $|\Psi_f\rangle$ represent exact initial and final states, respectively, in which the states consist of quantum-dot states, photon fields, and some external degrees of freedom, such as a substrate and a glass fiber probe. Since the exact states cannot be given rigorously, we deal with the problem for taking the minimum matter and photon states by using the projection operator method, where the theoretical treatment in such complex system comes down to two-body problem as we have reported in detail [16, 20].

We can rewrite the exact eigenstate as two substates which belong in a relevant P-space and an irrelevant Q-space, which are expressed by using projection operators P and Q as $|\Psi_\lambda^P\rangle = P|\Psi_\lambda\rangle$ and $|\Psi_\lambda^Q\rangle = Q|\Psi_\lambda\rangle$, respectively, where $\lambda = i, f$. Here, P and Q are specified by the following relations: $P + Q = 1$, $P^2 = P$, $Q^2 = Q$, $P^\dagger = P$, and $Q^\dagger = Q$ [21]. In the case of two-quantum-dot system, P-space is constructed from the eigenstates of $\widehat{H}_0$, i.e., a combination of the two energy levels for each quantum dot and the exciton–polariton vacuum state. In Q-space, which is complementary to P-space, the exciton–polariton states are not vacant. According to this notation, the exact state can be formally expressed by using the state in P-space only as

$$|\Psi_\lambda\rangle = \hat{J}P(P\hat{J}^\dagger \hat{J}P)^{-1/2}|\Psi_\lambda^P\rangle , \tag{30}$$

where

$$\hat{J} = \left[1 - (E_\lambda - \widehat{H}_0)^{-1}Q\widehat{H}_{int}\right]^{-1} , \tag{31}$$

and E_λ represents the eigenenergy of the total Hamiltonian $\widehat{H}$. Using (30), we can obtain the effective interaction $\widehat{H}_{eff}$ as

$$\langle \Psi_f | \widehat{H}_{int} | \Psi_i \rangle = \langle \Psi_f^P | \widehat{H}_{eff} | \Psi_i^P \rangle , \tag{32}$$

where

$$\widehat{H}_{eff} = (P\hat{J}^\dagger \hat{J}P)^{-1/2}(P\hat{J}^\dagger \widehat{H}_{int} \hat{J}P)(P\hat{J}^\dagger \hat{J}P)^{-1/2} . \tag{33}$$

To evaluate further (32), we approximate operator $\hat{J}$ and eigenvalue E_λ perturbativelly with respect to $\widehat{H}_{int}$; that is, $\hat{J} = 1 + (E_0^P - E_0^Q)^{-1}\widehat{H}_{int} + \cdots$. Since the lowest order is $\langle \Psi_f^P | P\widehat{H}_{int}P | \Psi_i^P \rangle = 0$, (32) is rewritten within the second order as

$$\hbar U = \sum_m \left\langle \Psi_f^P | \hat{H}_{\text{int}} | m^Q \right\rangle \left\langle m^Q | \hat{H}_{\text{int}} | \Psi_i^P \right\rangle \left(\frac{1}{E_{0i}^P - E_{0m}^Q} + \frac{1}{E_{0f}^P - E_{0m}^Q} \right) , \quad (34)$$

where E_{0i}^P, E_{0f}^P, and E_{0m}^Q represent the eigenenergies of the unperturbed Hamiltonian for the initial and final states in P-space and the intermediate state in Q-space, respectively. Since we focus on the interdot interaction of (34), we set the initial and final states in P-space to $|\Psi_i^P\rangle = |\Phi_{m\mu}^A\rangle|\Phi_g^B\rangle|0\rangle$ and $|\Psi_f^P\rangle = |\Phi_g^A\rangle|\Phi_{m'\mu'}^B\rangle|0\rangle$. Then, the intermediate states in Q-space that involve exciton–polaritons with the wavevector $\boldsymbol{k}$ are utilized for the energy transfer from one quantum dot to the other, according to $|m^Q\rangle = |\Phi_g^A\rangle|\Phi_g^B\rangle|\boldsymbol{k}\rangle$ and $|\Phi_{m\mu}^A\rangle|\Phi_{m'\mu'}^B\rangle|\boldsymbol{k}\rangle$. The superscripts A and B are used to label two quantum dots. Substituting (28), one can rewrite (34) as

$$\hbar U = \varphi_\mu^A(0)\varphi_{\mu'}^{B*}(0) \iint F_m^A(\boldsymbol{R}_A)F_{m'}^{B*}(\boldsymbol{R}_B)$$
$$\times (Y_A(\boldsymbol{R}_A - \boldsymbol{R}_B) + Y_B(\boldsymbol{R}_A - \boldsymbol{R}_B))\mathrm{d}\boldsymbol{R}_A\mathrm{d}\boldsymbol{R}_B , \quad (35)$$

where the sum of $\boldsymbol{R}_\alpha$ ($\alpha = \mathrm{A, B}$) in (28) is transformed to the integral form. The functions $Y_\alpha(\boldsymbol{R}_{AB})$, which connect the two spatially isolated two envelope functions $F_m^A(\boldsymbol{R}_A)$ and $F_m^B(\boldsymbol{R}_B)$, are defined as

$$Y_\alpha(\boldsymbol{R}_{AB}) = -\frac{1}{4\pi^2} \sum_{\lambda=1}^2 \int (\boldsymbol{\mu}_{cv}^A \cdot \hat{\boldsymbol{e}}_\lambda(\boldsymbol{k}))(\boldsymbol{\mu}_{cv}^B \cdot \hat{\boldsymbol{e}}_\lambda(\boldsymbol{k}))f^2(k)$$
$$\times \left(\frac{e^{i\boldsymbol{k}\cdot\boldsymbol{R}_{AB}}}{E(k) + E_\alpha} + \frac{e^{-i\boldsymbol{k}\cdot\boldsymbol{R}_{AB}}}{E(k) - E_\alpha} \right) \mathrm{d}\boldsymbol{k} , \quad (36)$$

where $\boldsymbol{R}_{AB} = \boldsymbol{R}_A - \boldsymbol{R}_B$ is used.

In order to obtain an explicit functional form of $Y_\alpha(\boldsymbol{R}_{AB})$, we apply the effective mass approximation to the exciton–polaritons

$$E(k) = \frac{\hbar^2 k^2}{2m_\mathrm{p}} + E_\mathrm{m} , \quad (37)$$

where m_p is the exciton–polariton effective mass. Using this approximation, (36) can be transformed into

$$Y_\alpha(\boldsymbol{R}_{AB}) =$$
$$(\boldsymbol{\mu}_{cv}^A \cdot \boldsymbol{\mu}_{cv}^B)\left[W_{\alpha+}e^{-\Delta_{\alpha+}R_{AB}} \left(\frac{\Delta_{\alpha+}^2}{R_{AB}} + \frac{\Delta_{\alpha+}}{R_{AB}^2} + \frac{1}{R_{AB}^3} \right) \right.$$
$$\left. -W_{\alpha-}e^{-\Delta_{\alpha-}R_{AB}} \left(\frac{\Delta_{\alpha-}^2}{R_{AB}} + \frac{\Delta_{\alpha-}}{R_{AB}^2} + \frac{1}{R_{AB}^3} \right) \right]$$
$$-(\boldsymbol{\mu}_{cv}^A \cdot \hat{\boldsymbol{R}}_{AB})(\boldsymbol{\mu}_{cv}^B \cdot \hat{\boldsymbol{R}}_{AB})\left[W_{\alpha+}e^{-\Delta_{\alpha+}R_{AB}} \left(\frac{\Delta_{\alpha+}^2}{R_{AB}} + \frac{3\Delta_{\alpha+}}{R_{AB}^2} + \frac{3}{R_{AB}^3} \right) \right.$$
$$\left. -W_{\alpha-}e^{-\Delta_{\alpha-}R_{AB}} \left(\frac{\Delta_{\alpha-}^2}{R_{AB}} + \frac{3\Delta_{\alpha-}}{R_{AB}^2} + \frac{3}{R_{AB}^3} \right) \right] , \quad (38)$$

where R_{AB} and $\hat{\boldsymbol{R}}_{AB}$ are the absolute value $|\boldsymbol{R}_{AB}|$ and the unit vector defined by $\boldsymbol{R}_{AB}/R_{AB}$, respectively. The weight coefficients $W_{\alpha\pm}$ and decay constants $\Delta_{\alpha\pm}$ are given by

$$W_{\alpha\pm} = \frac{\sqrt{E_{\mathrm{p}}}}{E_{\alpha}} \frac{(E_{\mathrm{m}} - E_{\alpha})(E_{\mathrm{m}} + E_{\alpha})}{(E_{\mathrm{m}} - E_{\mathrm{p}} \mp E_{\alpha})(E_{\mathrm{m}} \pm E_{\alpha}) - E_{\mathrm{m}}^2/2} , \tag{39}$$

$$\Delta_{\alpha\pm} = \frac{1}{\hbar c} \sqrt{E_{\mathrm{p}}(E_{\mathrm{m}} \pm E_{\alpha})} , \tag{40}$$

where the exciton–polariton effective mass is rewritten as $E_{\mathrm{p}} = m_{\mathrm{p}}c^2$. Since the dipole moments $\boldsymbol{\mu}_{\mathrm{cv}}^{\mathrm{A}}$ and $\boldsymbol{\mu}_{\mathrm{cv}}^{\mathrm{B}}$ are not determined as fixed values, we assume that they are parallel, and take a rotational average of (38). Therefore, $\langle(\boldsymbol{\mu}_{\mathrm{cv}}^{\mathrm{A}} \cdot \hat{\boldsymbol{R}}_{AB})(\boldsymbol{\mu}_{\mathrm{cv}}^{\mathrm{B}} \cdot \hat{\boldsymbol{R}}_{AB})\rangle = \mu_{\mathrm{cv}}^{\mathrm{A}}\mu_{\mathrm{cv}}^{\mathrm{B}}/3$ with $\mu_{\mathrm{cv}}^{\alpha} = |\boldsymbol{\mu}_{\mathrm{cv}}^{\alpha}|$, and we obtain the final form of the function $Y_{\alpha}(R_{AB})$ as

$$Y_{\alpha}(R_{AB}) = \frac{2\mu_{\mathrm{cv}}^{\mathrm{A}}\mu_{\mathrm{cv}}^{\mathrm{B}}}{3R_{AB}} \left(W_{\alpha+}\Delta_{\alpha+}^2 e^{-\Delta_{\alpha+}R_{AB}} - W_{\alpha-}\Delta_{\alpha-}^2 e^{-\Delta_{\alpha-}R_{AB}} \right) . \tag{41}$$

Equation (41) is the sum of two Yukawa functions with a short and long interaction range (heavy and light effective mass) given in (40). We can estimate the coupling strength between two quantum dots from the analytic form of the interaction potential given by (35) and (41), and we can show the existence of dipole-forbidden energy transfer driven by the optical near-field coupling, as discussed in the following.

Numerical Results

In this section, we give typical values of the coupling strength of $\hbar U$ in (35) using an example of CuCl quantum cubes embedded in an NaCl matrix. Due to the effect of size confinement, the center of mass motion and relative motion for an exciton in a CuCl quantum cube are [19]

$$F_{\boldsymbol{m}}^{\alpha}(\boldsymbol{R}_{\alpha}) = \left(\frac{2}{L_{\alpha}}\right)^{3/2} \sin\left(\frac{\pi m_x x_{\alpha}}{L_{\alpha}}\right) \sin\left(\frac{\pi m_y y_{\alpha}}{L_{\alpha}}\right) \sin\left(\frac{\pi m_z z_{\alpha}}{L_{\alpha}}\right) , \tag{42}$$

$$\varphi_{1s}(r) = \frac{1}{\sqrt{\pi a^3}} e^{-r/a} , \tag{43}$$

respectively, where the atomic site and the quantum number are represented by $\boldsymbol{R}_{\alpha} = (x_{\alpha}, y_{\alpha}, z_{\alpha})$ with $\alpha = $ A, B and $\boldsymbol{m} = (m_x, m_y, m_z)$ with $m_x, m_y, m_z = 1, 2, 3, \cdots$. The variables L_{α} and a denote a width of the quantum cube and the Bohr radius of the exciton, respectively. Here, we assume relative motion in the 1s state. The coupling strength is obtained numerically by substituting (41) and (42) into (35). In Fig. 5a, the calculation results are plotted as a function of the intercube distance. The curve with square dots represents the coupling between the dipole-active exciton levels, i.e., $\boldsymbol{m} = \boldsymbol{m}' = (1, 1, 1)$, in two quantum cubes. When we set the intercube distance and a width of the quantum cubes as $d = 5\,\mathrm{nm}$ and $L_{\mathrm{A}} = L_{\mathrm{B}} = 10\,\mathrm{nm}$,

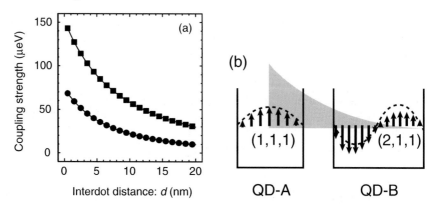

Fig. 5. (a) Coupling strength of the optical near field between pairs of CuCl quantum cubes embedded in an NaCl matrix. The curves shown with square and circular dots correspond to quantum numbers for the exciton center of mass motion $m = m' = (1,1,1)$, and $m = (1,1,1)$ and $m' = (2,1,1)$, respectively. The energy level $m' = (2,1,1)$ is a dipole-inactive state for conventional far-field light. The parameters are set as $E_A = E_B = 3.22\,\mathrm{eV}$, $E_m = 6.9\,\mathrm{eV}$, $\mu_{cv}^A = \mu_{cv}^B = 1.73 \times 10^{-2}\,\mathrm{eV}$, $L_A = 10\,\mathrm{nm}$, $L_B = 10$ and $14.1\,\mathrm{nm}$ ($m' = (1,1,1)$ and $(2,1,1)$), and $a = 0.67\,\mathrm{nm}$. (b) Schematic illustration of a transition between dipole-active and dipole-inactive states via the optical near-field coupling. Steeply gradient optical near field enables to excite near side local dipoles in a quantum dot with dipole-inactive $(2,1,1)$-level

respectively, the coupling strength is about $89\,\mu\mathrm{eV}$ ($U^{-1} = 7.4\,\mathrm{ps}$). The curve with circular dots is the result for $m = (1,1,1)$ and $m' = (2,1,1)$. For conventional far-field light, $m' = (2,1,1)$ is the dipole-inactive exciton level, and it follows that the optical near-field interaction inherently involves such a transition because of the finite interaction range. Figure 5b is a schematic illustration of the dipole-inactive transition, in which the optical near field enables to excite the local dipoles at the near side in a quantum dot with dipole-inactive level for far-field light. This coupling strength is estimated from Fig. 5a as $\hbar U = 37\,\mu\mathrm{eV}$ ($U^{-1} = 17.7\,\mathrm{ps}$) for $d = 5\,\mathrm{nm}$, and $\hbar U = 14\,\mu\mathrm{eV}$ ($U^{-1} = 46.9\,\mathrm{ps}$) for $d = 15\,\mathrm{nm}$, where the cube sizes are set as $L_A = 10\,\mathrm{nm}$ and $L_B = 14.1\,\mathrm{nm}$ to realize resonant energy transfer between the exciton state in QD-A and the first exciton excitation state in QD-B. The coupling strength ($m \neq m'$) is approximately half that of $m = m'$ at the same intercube distance, but it is strong enough for our proposed nanophotonic devices. For functional operations, the difference between the coupling strengths is important to divide the system into two parts, i.e., a quantum mechanical part and a classical dissipative part, as illustrated in Fig. 4.

2.4 Summary

In this section, we formulate a optical near-field coupling by using appropriate bases which are constructed form typical excitonic states in a quantum dot and exciton–polariton state in a surrounding system, and not using the long wave approximation which often applies to a conventional optical interaction in an atomic system. Although we have derived the coupling in the lowest order as given in (34), our formulation would be exact if we take rigorous eigenstate of exciton–polaritons as the intermediate states, instead of the simple effective mass approximation which is applied in the above discussion. However, in the following sections, our interests are characteristic functional operations of nanophotonic devices on the basis of certain coupling strength of the optical near field, rather than to understand fundamental properties of optical near-field coupling. More rigorous description of the optical near-field coupling will discuss elsewhere.

From numerical results shown in Fig. 5, the coupling strength of optical near field depends on the interdot distance, which is one of key features for nanophotonic device operations. By using this, we can control the dynamics of energy flow in nanometric space and develop some functional operations inherent to nanophotonic devices. Furthermore, we showed that dipole inactive energy transfer can occur when a distance between isolated quantum systems becomes enough small, which is related to the energy states in nanometric objects as well as steeply gradient spatial distribution of the optical near field. Especially, the dipole-inactive energy transfer between the states with different quantum numbers enables to realize unidirectional energy transfer in a nanometric system with the help of fast relaxation of exciton sublevels. This is a quite important feature for signal isolation in nanophotonic devices. In Sects. 3 and 4, we discuss operation principles of various functional devices by using such features of the optical near-field coupling skillfully.

3 Nanophotonic Switch Based on Dissipation Control

In Sect. 2, we had theoretically explained that an exciton in a dipole-inactive energy level can be excited by using an optical near field. A relaxation time of the exciton in the dipole-inactive level, the higher energy sublevel, is generally in the order of a few ps because of the strong coupling between an exciton and a phonon reservoir in a surrounding system [22]. Since the coupling strength of the optical near field corresponds to about subhundred ps, which has been estimated in Sect. 2, the intra-sublevel relaxation is as a figure fast as in the order of energy transfer between two quantum dots. Therefore, unidirectional energy transfer can be realized in a two or more quantum-dot system by mediating the intra-sublevel relaxation. On the other hand, we can create and annihilate an exciton in an exciton ground state by using external pumping light. Excitons in a quantum-dot system affect exciton–exciton interaction in a quantum dot, because more than an exciton confined to nanometric space. We have qualitatively regarded the excitons as fermionic particles, that is of course exact. When the lowest energy sublevel is occupied, the exciton population cannot drop into the lowest energy level,

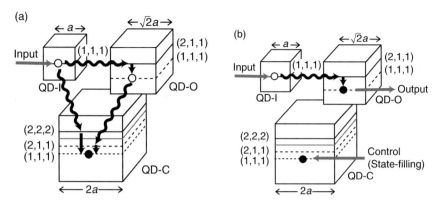

Fig. 6. A nanophotonic switch consisting of three quantum cubes with discrete energy levels showing the (**a**) OFF- and (**b**) ON-states

and thus, we can change the dissipation path selectively by arranging several quantum dots. This selectivity reads the origin of a nanophotonic switching operation.

In this section, we investigate our proposed nanophotonic switch, which is a basic element of nanophotonic devices [23]. Figure 6 illustrates a switch that consists of three quantum dots (cubes) with discrete exciton energy levels depending on the quantum-dot size. The one-side lengths of these cubes are chosen in the ratio $1:\sqrt{2}:2$, so that the adjacent quantum dots have resonant energy levels. The principle of operation of the switch is as follows: as shown in Fig. 6a, an exciton or population is created at the $(1,1,1)$-level in QD-I as an initial condition. Then the population is transferred to QD-O and QD-C as a result of an optical near-field coupling. Owing to the fast relaxation between sublevels in each dot via exciton–phonon coupling, the population is transferred to lower energy levels, and finally collected at the lowest $(1,1,1)$-level in QD-C. This corresponds to the OFF-state of the switch, and, consequently, we obtain no output signals from the output port, i.e., the $(1,1,1)$-level in QD-O. By contrast, in the ON-state of the switch (Fig. 6b), the $(1,1,1)$-level in QD-C is initially filled by the control light, isolating QD-C from the other two quantum dots. The input population only reaches the $(1,1,1)$-level in QD-O and can be detected as output signals, either by the optical near-field coupling to the detector or by far-field light emitted with electron–hole recombination.

From the above explanation, we understand that the key parameters determining the response time of the device are the coupling strength between two quantum dots via an optical near fields, and that between excitons and a phonon reservoir. In Sect. 3.1, dynamics of exciton population is formulated on the basis of quantum mechanical density matrix formalism, where we consider the phonon field as well as the optical near field discussed in Sect. 2, and roles of some key parameters in such a quantum-dot system are numerically clarified. This allows us to discuss the temporal dynamics of our proposed nanophotonic. We evaluate the response time of the CuCl quantum-cube system as a numerical example, which have been extensively examined in experimental and theoretical studies of quantum dots [19, 22, 24, 25]. Section 3.2 devotes to evaluate switching operations in a three quantum-dot system as shown in Fig. 6, where the effect of state-filling is introduced phenomenologically. Furthermore, faster iterative switching operations can be achieved in the order of 100 ps, when we apply appropriate control light pulse for utilizing stimulated absorption and emission effectively, which will be discussed by means of numerical analysis in Sect. 3.3.

3.1 Dynamics in a Two-Quantum-Dot System with Dissipation

As mentioned above, relaxation in the exciton sublevels guarantees the unidirectional energy transfer in a system with several quantum dots. The relaxation originates from coupling between exciton excited state and lattice vibrations in a quantum dot and surrounding matter which are regarded as a phonon reservoir. In order to understand energy transfer dynamics in such a quantum-dot system, which goes through a dissipative process, we first examine a two-quantum-dot system coupled to the phonon reservoir.

Formulation

In Fig. 7, we schematically illustrate a considered two-quantum-dot system and a phonon reservoir system, in which all energy transfer paths are depicted except for the coupling to far-field light because of different time scales. The Hamiltonian of the system is modeled as

$$\widehat{H} = \widehat{H}_0 + \widehat{H}_{\mathrm{int}} + \widehat{H}_{\mathrm{SR}} \tag{44}$$

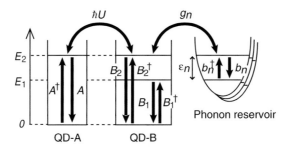

Fig. 7. Two-quantum-dot system. QD-A and QD-B are resonantly coupled due to an optical near-field interaction, and the sublevels in QD-B are coupled with the phonon reservoir

and

$$\widehat{H}_0 = \hbar\Omega_2 \hat{A}^\dagger \hat{A} + \hbar\Omega_1 \widehat{B}_1^\dagger \widehat{B}_1 + \hbar\Omega_2 \widehat{B}_2^\dagger \widehat{B}_2 + \hbar\sum_n \omega_n \hat{b}_n^\dagger \hat{b}_n \,, \qquad (45)$$

$$\widehat{H}_{\mathrm{int}} = \hbar U (\hat{A}^\dagger \widehat{B}_2 + \widehat{B}_2^\dagger \hat{A}) \,, \qquad (46)$$

$$\widehat{H}_{\mathrm{SR}} = \hbar\sum_n (g_n \hat{b}_n^\dagger \widehat{B}_1^\dagger \widehat{B}_2 + g_n^* \hat{b}_n \widehat{B}_2^\dagger \widehat{B}_1) \,. \qquad (47)$$

When we assume that initial and final states are constructed only in terms of one-exciton states, the creation (annihilation) operators of excitons can be written as follows: $\hat{A}^\dagger = [|e\rangle\langle g|]_A$ $(\hat{A} = [|g\rangle\langle e|]_A)$, $\widehat{B}_1^\dagger = [|e\rangle\langle g|]_{B_1}$, $(\widehat{B}_1 = [|g\rangle\langle e|]_{B_1})$, and $\widehat{B}_2^\dagger = [|e\rangle\langle g|]_{B_2}$, $(\widehat{B}_2 = [|g\rangle\langle e|]_{B_2})$. We can easily understand the following commutation relations: $[\widehat{B}_i^\dagger, \widehat{B}_j] = \delta_{i,j}([|e\rangle\langle e|]_{B_i} - [|g\rangle\langle g|]_B)$ and $[\widehat{B}_i, \widehat{B}_j] = [\widehat{B}_i^\dagger, \widehat{B}_j^\dagger] = 0$ $(i, j = 1, 2)$. Therefore, the operators are neither bosonic nor fermionic. Bosonic operators $(\hat{b}_n^\dagger, \hat{b}_n)$ are for the phonons with eigenenergy $\hbar\omega_n$. For simplicity, the rotating wave approximation is used in the interaction Hamiltonian $\widehat{H}_{\mathrm{int}}$ as $(\hat{A}+\hat{A}^\dagger)(\widehat{B}_2+\widehat{B}_2^\dagger) \approx \hat{A}^\dagger \widehat{B}_2 + \hat{A}\widehat{B}_2^\dagger$. Phonon reservoir is assumed to be a collection of multiple harmonic oscillators labeled n. Note that the exciton–polariton degrees of freedom have already been traced out, and thus the coupling strength of the optical near field, $\hbar U$, appears in (46). Dynamics of an exciton in this system is given by the following Liouville equation [26, 27]

$$\dot{\rho}(t) = -\frac{i}{\hbar}[\widehat{H}, \rho(t)] \,, \qquad (48)$$

where $\rho(t)$ represents the density operator, traced out the exciton–polariton degrees of freedom. In order to express the second-order temporal correlation clearly, the formal solution of (48) in the integral form is again substituted into the right-hand side of (48), and thus

$$\dot{\rho}^{\mathrm{I}}(t) = -\frac{i}{\hbar}\left[\widehat{H}_{\mathrm{int}} + \widehat{H}_{\mathrm{SR}}^{\mathrm{I}}(t), \hat{\rho}^{\mathrm{I}}(0)\right]$$
$$-\frac{1}{\hbar^2}\int_0^t \left[\widehat{H}_{\mathrm{int}} + \widehat{H}_{\mathrm{SR}}^{\mathrm{I}}(t), \left[\widehat{H}_{\mathrm{int}} + \widehat{H}_{\mathrm{SR}}^{\mathrm{I}}(t'), \hat{\rho}^{\mathrm{I}}(t')\right]\right] \mathrm{d}t' \,, \qquad (49)$$

where the superscript I means the interaction picture, and the relation $\widehat{H}_{\mathrm{int}}^{\mathrm{I}}(t) = \widehat{H}_{\mathrm{int}}$ is used [26]. Since we are interested in the exciton population in the two-quantum-dot system, we take a trace with respect to the degrees of freedom of the phonon reservoir as $\hat{\rho}_{\mathrm{S}}^{\mathrm{I}}(t) = \mathrm{Tr}_{\mathrm{R}}[\hat{\rho}^{\mathrm{I}}(t)]$. Here, the density operator is assumed to be a direct product of the quantum-dot system part $\hat{\rho}_{\mathrm{S}}^{\mathrm{I}}(t)$ and the reservoir system part $\hat{\rho}_{\mathrm{R}}^{\mathrm{I}}(t)$. If the reservoir has a very large volume, deviation from the initial value can be neglected, and the density operator is approximated as

$$\hat{\rho}^{\mathrm{I}}(t) = \hat{\rho}_{\mathrm{S}}^{\mathrm{I}}(t)\hat{\rho}_{\mathrm{R}}^{\mathrm{I}}(t) \approx \hat{\rho}_{\mathrm{S}}^{\mathrm{I}}(t)\hat{\rho}_{\mathrm{R}}(0) \,, \qquad (50)$$

which corresponds to the Born approximation [26]. Taking a trace on both sides of (49) about the reservoir operator, we obtain

$$
\begin{aligned}
\dot{\rho}_{\mathrm{S}}^{\mathrm{I}}(t) = {} & -\mathrm{i}U(r)[\hat{A}^\dagger \widehat{B}_2 + \widehat{B}_2^\dagger \hat{A}] \\
& - \sum_n n(\omega_n, T) \left[(\{\hat{C}\hat{C}^\dagger, \hat{\rho}_{\mathrm{S}}^{\mathrm{I}}(t)\} - 2\hat{C}^\dagger \hat{\rho}_{\mathrm{S}}^{\mathrm{I}}(t)\hat{C}) \otimes \gamma_n^{\mathrm{r}}(t) \right. \\
& \left. \quad - \mathrm{i}[\hat{C}\hat{C}^\dagger, \hat{\rho}_{\mathrm{S}}^{\mathrm{I}}(t)] \otimes \gamma_n^{\mathrm{i}}(t) \right] \\
& - \sum_n [1 + n(\omega_n, T)] \left[(\{\hat{C}^\dagger\hat{C}, \hat{\rho}_{\mathrm{S}}^{\mathrm{I}}(t)\} - 2\hat{C}\hat{\rho}_{\mathrm{S}}^{\mathrm{I}}(t)\hat{C}^\dagger) \otimes \gamma_n^{\mathrm{r}}(t) \right. \\
& \left. \quad + \mathrm{i}[\hat{C}^\dagger\hat{C}, \hat{\rho}_{\mathrm{S}}^{\mathrm{I}}(t)] \otimes \gamma_n^{\mathrm{i}}(t) \right. , \tag{51}
\end{aligned}
$$

where the curly brackets $\{\cdot\}$ represent the anti-commutation relation, and the notation $\otimes$ designates the convolution integral. In order to avoid verbose expression, we make the following replacement: $\hat{C}^\dagger = \widehat{B}_2^\dagger\widehat{B}_1$ and $\hat{C} = \widehat{B}_1^\dagger\widehat{B}_2$. Since we assume that the reservoir system is at equilibrium, the terms including $\mathrm{Tr}_R[\hat{b}_n^\dagger \hat{\rho}_R(0)]$ and $\mathrm{Tr}_R[\hat{b}_n \hat{\rho}_R(0)]$ disappear in (51). The number of phonons in the equilibrium state is written as $n(\omega_n, T) = \mathrm{Tr}_R[\hat{b}_n^\dagger\hat{b}_n\hat{\rho}_R(0)]$, and it follows Bose–Einstein statistics as

$$
n(\omega_n, T) = \frac{1}{e^{\hbar\omega_n/k_{\mathrm{B}}T} - 1} . \tag{52}
$$

The real and imaginary parts of function

$$
\gamma_n(t) = |g_n|^2 e^{\mathrm{i}(\Delta\omega - \omega_n)t} \tag{53}
$$

with $\hbar\omega = \hbar(\Omega_2 - \Omega_1)$ are represented as $\gamma_n^{\mathrm{r}}(t)$ and $\gamma_n^{\mathrm{i}}(t)$, respectively, and are related to the relaxation (real part) and energy shift (imaginary part) of the energy level in QD-B that is originated from the coupling to the phonon reservoir. The convolution integral in (51) expresses a memory effect due to time delay in the phonon reservoir. However, if the dynamics of the reservoir system are much faster than those of the two-quantum-dot system, one can approximate the density operator of the two-dot system as $\hat{\rho}_{\mathrm{S}}^{\mathrm{I}}(t - t') = \hat{\rho}_{\mathrm{S}}^{\mathrm{I}}(t)$ (a Markov approximation). Using this approximation, and rewriting the summation as $\sum_n = \int_0^\infty D(\omega)\,\mathrm{d}\omega$, with $D(\omega)$ being the density of states for each phonon, we can express the convolution integral analytically as

$$
\begin{aligned}
& \sum_n n(\omega_n, T)\hat{\rho}_{\mathrm{S}}^{\mathrm{I}}(t) \otimes \gamma_n(t) \\
& = \hat{\rho}_{\mathrm{S}}^{\mathrm{I}}(t) \int_0^\infty n(\omega, T)D(\omega)|g(\omega)|^2 \left(\int_0^t e^{\mathrm{i}(\Delta\omega - \omega)t'}\,\mathrm{d}t' \right) \mathrm{d}\omega \\
& \approx \hat{\rho}_{\mathrm{S}}^{\mathrm{I}}(t) \left[\pi n(\Delta\omega, T)D(\Delta\omega)|g(\Delta\omega)|^2 \right. \\
& \left. \quad + \mathrm{i}P \int_0^\infty \frac{n(\omega, T)D(\omega)|g(\omega)|^2}{\Delta\omega - \omega}\,\mathrm{d}\omega \right] . \tag{54}
\end{aligned}
$$

Here, we extend the upper limit of the time integration to infinity. The equation of motion for the dot system is finally reduced to

$$\dot{\rho}_S^I(t) = iU(r)[\hat{A}^\dagger \hat{B}_2 + \hat{B}_2^\dagger \hat{A}, \hat{\rho}_S^I(t)] - n\gamma(\{\hat{C}\hat{C}^\dagger, \hat{\rho}_S^I(t)\} - 2\hat{C}^\dagger \hat{\rho}_S^I(t)\hat{C})$$
$$- (1+n)\gamma(\{\hat{C}^\dagger \hat{C}, \hat{\rho}_S^I(t)\} - 2\hat{C}\hat{\rho}_S^I(t)\hat{C}^\dagger) , \tag{55}$$

where $n \equiv n(\Delta\omega, T)$ and $\gamma \equiv \pi D(\Delta\omega)|g(\Delta\omega)|^2$. The terms indicating the energy shift are neglected in (55) because the shift is usually small in the case of weak coupling between the quantum-dot system and phonon reservoir.

Let us consider one-exciton dynamics in the system, using three bases, as illustrated in Fig. 8. The equations of motion for the matrix elements are then read in the Schrödinger picture as

$$\dot{\rho}_{11}(t) = iU(r)[\rho_{12}(t) - \rho_{21}(t)], \tag{56}$$

$$\dot{\rho}_{12}(t) - \dot{\rho}_{21}(t) = 2iU(r)[\rho_{11}(t) - \rho_{22}(t)] - (1+n)\gamma[\rho_{12}(t) - \rho_{21}(t)], \tag{57}$$

$$\dot{\rho}_{22}(t) = -iU(r)[\rho_{12}(t) - \rho_{21}(t)] - 2(1+n)\gamma\rho_{22}(t) + 2n\gamma\rho_{33}(t), \tag{58}$$

$$\dot{\rho}_{33}(t) = 2(1+n)\gamma\rho_{22}(t) - 2n\gamma\rho_{33}(t) , \tag{59}$$

where $\rho_{mn}(t) \equiv \langle \Phi_m|\hat{\rho}_S(t)|\Phi_n\rangle$ is employed. When the temperature, T, equals zero $(n = 0)$, (56–59) can be solved analytically. The diagonal parts representing the population probability for each energy level in QD-A and QD-B, as well as the off-diagonal parts representing quantum coherence, are given as

$$\rho_{11}(t) = \frac{1}{Z^2}e^{-\gamma t}\left[\frac{\gamma}{2}\sinh(Zt) + Z\cosh(Zt)\right]^2 , \tag{60}$$

$$\rho_{22}(t) = \frac{U^2}{Z^2}e^{-\gamma t}\sinh^2(zt) , \tag{61}$$

$$\rho_{33}(t) = 1 - [\rho_{11}(t) + \rho_{22}(t)] , \tag{62}$$

$$\rho_{12}(t) = -\rho_{21}(t) = i\frac{U}{Z^2}e^{-\gamma t}\sinh(Zt)\left[\frac{\gamma}{2}\sinh(Zt) + Z\cosh(Zt)\right] , \tag{63}$$

where $Z \equiv \sqrt{(\gamma/2)^2 - U^2}$, and initial conditions $\rho_{11}(0) = 1$ and $\rho_{12}(0) = \rho_{21}(0) = \rho_{22}(0) = \rho_{33}(0) = 0$ are used. We define the state-filling time τ_S as $\rho_{33}(\tau_S) = 1 - e^{-1}$, which corresponds to the time for the excitation energy

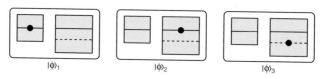

$|\phi\rangle_1$ $\qquad\qquad$ $|\phi\rangle_2$ $\qquad\qquad$ $|\phi\rangle_3$

Fig. 8. Three bases of the single-exciton state in a two-quantum-dot system

transfer from QD-A to the lower energy level in QD-B. From (60–68), it follows that the temporal evolution of the population is quite different at $U > \gamma/2$ and $U < \gamma/2$. Although (60–68) seems to be undefined at $U = \gamma/2$ ($Z = 0$), taking a limit value, there is a definite solution regardless of whether $Z \to +0$ or -0 is taken. In Fig. 9, the state-filling time τ_S is plotted as a function of the ratio of $\gamma/2$ to U. For $U > \gamma/2$, population shows damped oscillation with envelope function $e^{-\gamma t}$; thus, τ_S is determined by the relaxation constant γ, i.e., $\tau_S \sim \gamma^{-1}$. By contrast, for $U < \gamma/2$, $\rho_{22}(t)$ decays monotonically. At first glance, as $\gamma/2$ increases, we expect τ_S to decrease monotonically, because the population flows into the lower energy level more quickly; nevertheless, τ_S increases again, as shown in Fig. 9. This occurs because the upper energy level in QD-B becomes effectively broad with increasing γ, which results in a decrease in the resonant energy transfer between the quantum dots. When the ratio $\gamma/2U$ is sufficiently large, τ_S increases linearly, as seen in Fig. 9. Therefore, the state-filling time is not only determined by the coupling strength between two quantum dots via the optical near field, but also by the coupling strength to the phonon reservoir system. From Fig. 9, it follows that the fastest energy transfer is obtained when $\gamma/2 \sim U$ is satisfied.

The term $2\gamma n\rho_{33}(t)$ on the right-hand side of (58) indicates that the finite temperature effect due to the finite number of phonons ($n \neq 0$) induces back transfer of the excitation energy from the reservoir to the two-quantum-dot system. Within the Born approximation adopted in (50), this term incoherently increases population $\rho_{22}(t)$. As population $\rho_{33}(t)$ increases, the back transfer becomes large, and gives residual populations $\rho_{11}(t)$ and $\rho_{22}(t)$ in the upper-levels in both quantum dots.

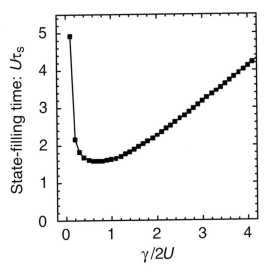

Fig. 9. The state-filling time τ_S plotted as a function of the ratio of the exciton–phonon coupling strength $\gamma/2$ to the interdot optical near-field coupling strength U

Numerical Results

Using (56–63), we present numerical results for the dynamics of exciton population for zero and finite temperatures in order to verify the theoretical consideration just presented, and to estimate the state-filling time for some practical cases. Suppose that the system consists of two CuCl quantum cubes embedded in an NaCl matrix. The quantum cube size is set as $a:\sqrt{2}a$, so that the $(1,1,1)$-level of QD-A is resonant with the $(2,1,1)$-level of QD-B. The variable parameters are the coupling strength $\hbar U$ or the intercube distance, the relaxation constant γ, and the temperature T of the phonon reservoir system.

In Fig. 10a and b, the E_2 and E_1-level populations $\rho_{22}(t)$ and $\rho_{33}(t)$ in QD-B are plotted for the coupling strengths of the optical near field, $\hbar U = 100$, 60, and 40 μeV ($U = 1.51 \times 10^{-1}$, 0.90×10^{-1}, and 0.60×10^{-1} ps^{-1}). In order to investigate a behavior of the dynamics around the critical condition $U = \gamma/2$, appropriate coupling strengths are chosen, though these are a little strong compared with the estimated values in Sect. 2. The relaxation constant γ is assumed to be $(1\,\mathrm{ps})^{-1}$ from the experimental study [22], and the temperature is $T = 0\,\mathrm{K}$. The exciton populations for all three cases in Fig. 10a show monotonic decays, because $U < \gamma/2$ is satisfied. From Fig. 10b, the state-filling time τ_S is estimated for the coupling strength $\hbar U = 100$, 60, and 40 μeV as 22, 60, and 140 ps, respectively. This indicates that an energy transfer time of less than 100 ps can be realized when the coupling strength is larger than about 50 μeV.

Figure 11 shows the result for $\gamma^{-1} = 10\,\mathrm{ps}$, which corresponds to weaker coupling between an exciton and the phonon reservoir. Damped oscillation due to nutation between two resonant levels is clearly seen for $\hbar U = 100\,\mu\mathrm{eV}$ in Fig. 11a, where $U > \gamma/2$ is satisfied. Although $U > \gamma/2$ is also satisfied for $\hbar U = 60$ and 40 μeV, we cannot observe the oscillation because of the

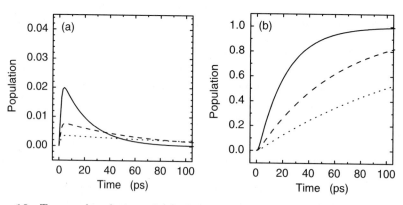

Fig. 10. Temporal-evolution of (**a**) E_2-level population $\rho_{22}(t)$ and (**b**) E_1-level population $\rho_{33}(t)$, where the relaxation constant γ and temperature T are set as $(1\,\mathrm{ps})^{-1}$ and 0 K, respectively. The *solid*, *dashed*, and *dotted curves* represent the cases for coupling strengths, $\hbar U = 100$, 60, and 40 μeV, respectively

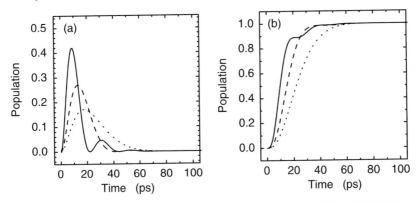

Fig. 11. Temporal-evolution of (**a**) E_2-level population $\rho_{22}(t)$ and (**b**) E_1-level population $\rho_{33}(t)$, where the relaxation time γ and temperature are set as $(10\,\text{ps})^{-1}$ and $0\,\text{K}$, respectively. The *solid, dashed,* and *dotted curves* represent the cases for coupling strengths, $\hbar U = 100$ 60, and $40\,\mu\text{eV}$, respectively

small amplitudes. From Fig. 11b, the state-filling time τ_S for the three cases is estimated as 12, 18, and 25 ps, respectively. Compared to Fig. 10b, the state-filling speed becomes faster in Fig. 11b in spite of the decrease in γ because U is nearly equal to $\gamma/2$. Figures 10 and 11 indicate that the intercube distance should be adjusted so that the optical near-field coupling is of the same order as the exciton–phonon coupling in order to obtain the fastest energy transfer in the system.

Figure 12a and b shows the temperature dependence of populations $\rho_{22}(t)$ and $\rho_{33}(t)$, respectively. The temperature is set as either $T = 50$ or $100\,\text{K}$. These results are obtained by using a Laplace transform of (56–59), where singular points are derived numerically for the given numerical parameters. As mentioned in the beginning in Sect. 3, a finite temperature induces the incoherent back transfer of energy, and this results in residual populations at the upper energy levels in both quantum cubes. Figure 12a shows that the E_2-level population in QD-B converges on a finite temperature-dependent value that can be derived numerically and is denoted by the horizontal dotted lines in Fig. 12a. As shown in Fig. 12b, however, the time evolution of the E_1-level population is almost independent of the temperature, except for a decrease in the amplitude because of the residual populations at upper energy levels.

System Tolerance and Further Discussion

So far, the theoretical modeling of the population dynamics in a two-quantum-dot system has assumed an ideally perfect resonance condition, which may be too tight to fabricate such a system with definite size ratios. In order to release the resonance condition, we estimate allowable tolerance, or size deviation of quantum dots from designed values. When the deviation from

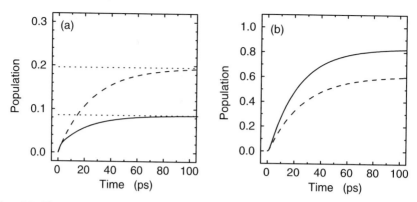

Fig. 12. Temperature dependence of (a) E_2-level population $\rho_{22}(t)$ and (b) E_1-level population $\rho_{33}(t)$, when $\hbar U = 100\,\mu\text{eV}$ and $\gamma = (1\,\text{ps})^{-1}$. The *solid* and *dashed curves* represent the cases for temperature $T = 50$ and $100\,\text{K}$, respectively. The two *dotted lines* denote the limit values of E_1-level population for both temperature at infinite time

the resonant energy $\hbar\Delta\Omega$ in QD-B is introduced, the factor on the right-hand side in (61) is modified, and the ratio of that in off-resonance to on-resonance is approximately proportional to $\gamma^2/(\gamma^2+\Delta\Omega^2)$ $(\gamma/2 \gg U)$. Therefore, we can achieve efficiency more than 50% even if the deviation $\Delta\Omega < \gamma$ is introduced. When the dot size and relaxation constant are set as $7.1\,\text{nm}$ $(= 5\sqrt{2}\,\text{nm})$ and $\hbar\gamma = 3\,\text{meV}$, respectively, approximately 10%-deviation of the dot size can be allowed. As the size of quantum dots is larger, the tolerable deviation is more relaxed. It is feasible to make such quantum dots by the recent advancement of nanofabrication techniques [28, 29]. In fact, experimental results in our research group [30] show the consistent population dynamics as we discussed above.

Our results and discussion might be valid within the Born–Markov approximation. Even if the Born approximation is admitted because of the large volume of the phonon reservoir, the Markov approximation may not be guaranteed at low temperature, as assumed in the transformation in (54). Therefore, we might need to deal with it more carefully, since it has been pointed out that non-Markov behavior manifests itself at low temperature [26,31]. The effects of non-Markov behavior on the dynamics are now under investigation.

We focused our attention on the energy transfer process in the nanometric regime, but it is quite important to investigate signal control, and to manipulate the electronic states of the components in a nanophotonic device, where we must deal with a two-exciton state, as shown in Fig. 6b. In the following, we show numerical results of exciton population dynamics with signal manipulation in a three-quantum-dot system, which is formulated by the same manner except for phenomenological insertion of control-light pulse.

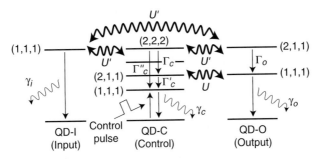

Fig. 13. Theoretical model of a nanophotonic switch

3.2 Nanophotonic Switch

Switching Operations

In Fig. 13, we show a theoretical model of the nanophotonic switch, in which each neighboring pair of quantum dots has a resonant energy level that is determined by setting the size as 1(I):2(C):$\sqrt{2}$(O), and thus, the system consists of two-level, three-level, and five-level quantum dots depending on the size ratios. In Sect. 3.1, we ignore radiative relaxation in the exciton ground levels because the radiative relaxation time is the order of 1 ns which is enough longer than the nonradiative one. However, for an iterative switching operation discussed here, exciton population has to be swept out to the external field and return into the initial state. This radiative relaxation, i.e., spontaneous emission of photons, is described in the same manner in (55) based on the Born–Markov approximation for a surrounding photon reservoir, where the radiative relaxation constants are much smaller than the nonradiative ones. In Fig. 13, the notations of radiative and nonradiative relaxation constants are distinguished by the Greek characters γ and Γ, respectively, where the subscripts of the relaxation constants represent the terminal quantum dots for the input, output, and control. For simplicity, we assume that both the photon and phonon reservoirs are empty ($T = 0$), where the effect of temperature is just lowering the signal contrast and does not affect the dynamics in the nanophotonic switch. In a switching device, we require another important element, that is injection of external control signal to change the ON- and OFF-states. Rigorous modeling of the external exciton excitation is quite difficult since matter coherence presents. Here, we assume weak excitation in which optical nutation or Rabi oscillation due to the control pulse does not appear, and approximately regards the excitation as an incoherent process. Thus, the external excitation can be also written by using the Born–Markov approximation with the very large number of photons. Adding the external excitation in (55), a phenomenological equation of motion for the density operator reads

$$\dot{\hat{\rho}}(t) = -\frac{i}{\hbar}[\widehat{H}_0 + \widehat{H}_{\text{int}}, \hat{\rho}(t)]$$
$$+ \text{(nonradiative relaxation part)} + \text{(radiative relaxation part)}$$
$$+ A_{\text{pump}}(t)(\hat{C}_1\hat{\rho}(t)\hat{C}_1^\dagger + \hat{C}_1^\dagger\hat{\rho}(t)\hat{C}_1)\,, \tag{64}$$

where $\hat{C}_1^\dagger$ and $\hat{C}_1$ represent the creation and annihilation operators of an exciton at the energy sublevel in QD-C, and the other verbose descriptions for the radiative and nonradiative relaxations are omitted. $A_{\text{pump}}(t) \propto n_{\text{ph}}$ (n_{ph}: photon number) is a time-dependent pumping rate of the control light pulse.

The operation of this switch is as follows. An input signal creates an exciton at the $(1, 1, 1)$-level in QD-I, and this transfers to the resonant energy levels of the adjacent two quantum dots via an optical near-field coupling [15]. While some energy levels even with total quantum numbers are optically (dipole) forbidden and are not directly excited by far-field light, the optical near field allows such transition because their spatial localization resolves the wave function of adjacent nanometric quantum dots far beyond light diffraction limits. The ON- or OFF-state of the switch corresponds to whether the lowest $(1, 1, 1)$-level in QD-C is unoccupied or occupied, respectively. When this energy level is occupied (state-filling), the initial excitation in QD-A ultimately reaches the $(2, 1, 1)$-level in QD-C or the $(1, 1, 1)$-level in QD-O, leading to an output signal. The lowest energy level in each quantum dot is coupled to a free photon reservoir to sweep out the excitation energy radiatively and to a laser photon reservoir to control this device by using a light pulse. Intra-sublevel transitions in QD-C and D guarantee unidirectional energy transfer in the-three quantum-dot system.

By using this model, we derived equations of motion for density matrix elements for possible one- and two-exciton states similar to (56), and calculated the exciton dynamics numerically. As an initial exciton population, a steady-state is prepared, where the $(1, 1, 1)$-level in QD-I is excited weakly and continuously. The output signal (luminescence intensity) is proportional to the $(1, 1, 1)$-level population in QD-O. Figure 14 shows the temporal evolution of the exciton population in QD-O after an incoherent short pulse (10 ps) is applied to QD-C. This situation corresponds to transition from the OFF-state to the ON-state. When state-filling is achieved instantaneously, the exciton population in QD-O increases and fast vibration appears as shown in Fig. 14. The vibration manifests itself only for the state-filling condition. This is apparently caused by nutation between the $(2, 1, 1)$ and $(1, 1, 1)$-levels in QD-C and QD-O, since the intra-sublevel transition in QD-C is prevented due to the state-filling. In this case, the switching speed from the OFF- to ON-states is about 2 ns, which is not so fast. However, as discussed in Sect. 3.1, we expect that the fastest switching time about sub-100 ps can be obtained, when the coupling strength of the optical near field is designed as the same order of nonradiative relaxation constant.

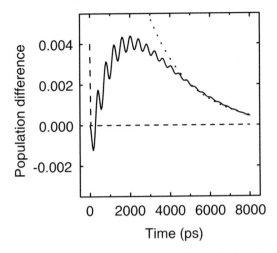

Fig. 14. Temporal evolution of the $(1, 1, 1)$-level population in QD-O from the initial steady-state. The adjustable parameters are set as follows: optical near-field coupling $U = U' = (130\,\text{ps})^{-1}$, radiative relaxation constants $\gamma_I = (16.8\,\text{ns})^{-1}$, $\gamma_C = (2.1\,\text{ns})^{-1}$, and $\gamma_O = (5.9\,\text{ns})^{-1}$, and nonradiative relaxation time $\Gamma_C = (20\,\text{ps})^{-1}$, $\Gamma_C'' = (10\,\text{ps})^{-1}$, $\Gamma_C'' = (30\,\text{ps})^{-1}$, and $\Gamma_O = (20\,\text{ps})^{-1}$

The readers may notice that the relaxation time in the later stage over 2 ns is very slow, which is approximately twice of an isolated quantum dot, and the switch seems to spend much time for recovery to the initial state. This long relaxation time is caused by the coherence between QD-C and QD-O during the state-filling in QD-C. Thus, the phenomenon cannot be reproduced using incoherent rate equation approach, and is a negative property in the nanophotonic switch. One way to realize fast recovery time is that another larger quantum dot which has fast relaxation time is located near QD-C to break the state-filling condition. We also have another interesting idea to use stimulated absorption and emission process for the recovery of the system, which is discussed in Sect. "Operation Using Stimulated Absorption and Emission".

Operation Using Stimulated Absorption and Emission

For switching and recovery operations in this system, we have found an important feature related to spontaneous absorption and emission process. At first glance, it seems that the switching device does not return the initial OFF-state unless it spends very long time depending on the radiative lifetime as demonstrated in Fig. 14. However, the exciton population in the system actually decays less than that of the initial steady-state during control light injection, and reaches another stable state, where the transition time from the initial state to the second steady-state is much faster than the spontaneous emission lifetime as shown in Fig. 15. The effect can be explained by the stimulated absorption and emission in QD-C. For the control light injected, the

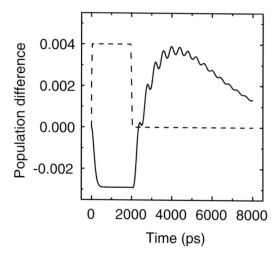

Fig. 15. Temporal evolution of the $(1, 1, 1)$-level population in QD-O from the initial steady-state. The applied control pulse width is set as 2 ns. The same adjustable parameters in Fig. 6 are adopted

exciton population in the lower energy sublevel in QD-C approaches by force a half of a unit value due to the stimulated absorption and emission. In the initial steady-state, the exciton population in the level is beyond a half value, and thus, the stimulated emission dominantly operates to decrease the population less than the initial steady-state, leading to the stable state with lower population. When we stop the control light injection, the exciton population increases beyond that of the beginning steady-state with the fast transition time depending on the coupling strength via the optical near field. Using these stimulated absorption and emission process, the fast and iterative switching operations can be achieved.

Figure 16 is one of the calculated results, where two control light pulses with 500 ps-pulse width are applied. We observe that the transition from ON- to OFF-states takes only a few 100 ps. With further optimization of the intercube distance, quantum-dot size, and surrounding materials of photon reservoir, we can obtain a switching time (OFF to ON, and ON to OFF) less than 100 ps, which is in the order of the inverse of optical near-field coupling strength, and is faster enough for highly integrated nanophotonic devices we proposed.

3.3 Summary

In this section, at first, we have formulated exciton dynamics in a two-quantum-dot system coupled to phonon reservoir, i.e., two-level and three-level quantum-dot systems, where energy transfer occurs between energy sublevels with different quantum numbers. Such energy transfer is inactive for

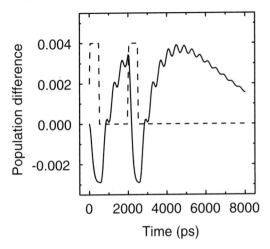

Fig. 16. Demonstration of the switching operation using two control light pulses with 500 ps-pulse width. The *solid* and *dashed curves* represent the population at the $(1, 1, 1)$-level in QD-O and the temporal profile of the control light pulses, respectively. The same adjustable parameters in Fig. 6 are adopted

far-field light, and is not realized unless using the optical near-field interaction. Moreover, owing to the fast relaxation via exciton–phonon coupling, unidirectional energy transfer can be realized in nanometric space, which is difficult in conventional optical devices when miniaturization of the device progresses. With the help of density operator formalism, temporal evolution can be solved analytically for zero temperature. As a result, we have found that an optimal condition exists for the fast switching operation; the coupling strength of an optical near field is comparable to the intra-sublevel relaxation constant. On the other hand, for finite temperature, the excitation from lower sublevels to upper sublevels occurs, and the signal contrast in this switch becomes lower. In order to improve the signal contrast, a mechanism to sweep out the population compulsorily from the output energy level is required. Although we have not mentioned in Sect. 3.1, stimulated absorption and emission process may be one of the useful phenomena for an improvement of the signal contrast, which has been described as another viewpoint in Sect. 3.2.

A fundamental switch is generally a three-terminal device for input, output, and control signals. We have investigated a three-quantum-dot system numerically as a nanophotonic switch. The energy transfer dynamics is almost same as a two-quantum-dot system, except for control excitation. When the control light pulse is applied, there are two excitons in the switch, and energy transfer path of an exciton changes due to a state-filling effect. For iterative operation of the nanophotonic switch, a mechanism to sweep out the exciton population is again required. We can use spontaneous emission for this, but the spontaneous lifetime is not as fast as in the order of 1 ns. Here, we have found that the stimulated absorption and emission process are valuable for the fast iterative operations, because the time for sweeping out the exciton population is free from the spontaneous emission lifetime and is determined by the coupling strength of the optical near field. By using these process and optimizing the coupling strength of the optical near field, we have roughly estimated the switching time of sub-100 ps. Such a nanophotonic switch has been already demonstrated experimentally in our research group by using CuCl quantum cubes embedded in an NaCl matrix [30], where they searched for an appropriate quantum-dot trio with the ratios of $1:\sqrt{2}:2$ by means of a spectroscopic method.

With further progress in nanofabrication techniques, any quantum dots will be aligned at desired positions, and thus, a highly integrated optical information processing device which consists of the nanophotonic switch can be realized. Furthermore, novel type of devices, which does not operate in a conventional way, should be proposed in the next stage. Our imagined form of such nanophotonic functional devices are composed by a quantum mechanical information processing part as well as a classical dissipative information processing part. In Sect. 4, such type of nanophotonic devices, in which coherently coupled states due to several quantum dots are positively utilized, will be explained analytically. We expect that the reader will feel a large possibility in nanophotonics and future nanophotonic device technologies.

4 Nanophotonic Functional Devices Using Coherently Coupled States

As we mentioned in Sect. 1, characteristic coupled states via an optical near-field interaction can be generated in a system without dissipation, such as symmetric and anti-symmetric states. These states extend two or more quantum systems with matter coherence as explained in Sect. 1.2, and have different energies in relation with the coupling strength of the optical near field. When we prepare another quantum system which interacts with these states, the energy difference affects resonant conditions of energy transfer via the optical near field. Therefore, the coherently coupled states can be utilized to pick up some information signal selectively.

In this section, we propose some nanophotonic functional devices, which consist of several quantum dots coupled via an optical near field. Such devices have two key operation parts; one is a *coherent operation part* or a quantum mechanical operation part, and the other is a *dissipative output part* (see Fig. 4 in Sect. 1). As we mentioned in the earlier sections, unidirectional energy transfer is indispensable in any functional devices for identification of a final state or an output signal, which can be guaranteed by using intra-sublevel relaxation due to exciton–phonon coupling. The dissipative output part includes such unidirectional irreversibility with the help of the energy sublevels in quantum dots. On the other hand, in the coherent operation part, an exciton excitation exists at the resonant energy levels in several quantum dots mediating the optical near field, where the coherently coupled states survives for a short period of time, before the excitation decays in the dissipative output part [32]. This section focuses on taking full advantage of these coherent operation part and dissipative output parts to realize functional operations based on nanophotonic inherent features [33]. In the earlier part of this section, as a typical example, we consider a three-quantum-dot system illustrated in Fig. 17. In this system, two identical quantum dots (QD-A and B) are resonantly coupled with each other via an optical near field, that consists of the coherent operation part, and a third quantum dot with larger size than the other two corresponds to the dissipative output part.

Various authors have investigated the coupling properties and dynamics in a pair of quantum dots. For example, the energy shift due to exciton–exciton or Coulomb interactions between electrons and holes has been evaluated theoretically to process quantum information [34,35], and a controlled-NOT logic gate has been proposed using the energy shift [36]. In these studies, excitons or qubits were controlled by two-color laser pulses of far-field light. As a similar subject to this section, Quiroga and Johnson [37] theoretically discussed the dynamics in two and three-quantum-dot systems and presented a way to prepare both quantum Bell and Greenberger–Horne–Zeilinger entangled states, by using far-field light, which allows only global excitation of two and three quantum dots with spatially symmetric arrangement. By contrast, we

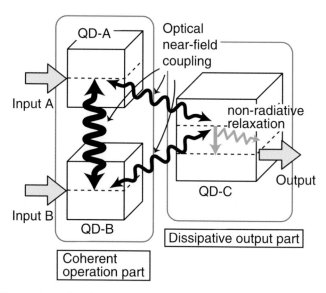

Fig. 17. Illustration of a three-quantum-dot system that consists of two identical two-level dots (QD-A and B) and a three-level dot (QD-C). Since the coupling between QD-A and B is stronger than that between QD-A and C (QD-B and C), the system is divided into two parts: a coherent operation part with optical nutation and a dissipative output part with nonradiative relaxation

deal with coupled quantum-dot systems arranged symmetrically and asymmetrically, which are individually excited by the optical near field, and the intra-sublevel relaxation is also considered for the unidirectional energy transfer. Note that the excitation in each quantum dot can be prepared individually owing to the spatial localization of the optical near field. The exciton dynamics driven by the optical near field has been investigated in the case of a coupled two-quantum-dot system with a relaxation process in Sect. 3.1. The energy transfer between two quantum dots is expressed as a Förster-like process [6], and the nutation of excitation occurs in the strongly coupled or resonant energy levels, corresponding to the coherent operation part in our system. For the short period before relaxation, certain coherently coupled states appear in the coherent operation part, depending on the initial excitation. In order to prepare the initial excitation, a shorter excitation time in the individual quantum dot than the energy transfer time between two identical quantum dots is necessary, where the excitation time is inversely proportional to the optical near-field intensity. The energy transfer time or coupling strength via an optical near field can be controlled by adjusting interdot distance. The population in the coherently coupled states can be transferred to the third quantum dot (QD-C) if the energy level of QD-C is adjusted to couple resonantly with the entangled states in the coherent operation part. If this happens, QD-C operates as the dissipative output part, which involves an intra-sublevel relaxation process due to the exciton–phonon interaction. In this manner, unidirectional energy or signal transfer is satisfied.

This section is organized as follows. First of all, we formulate the simplest case of three-quantum-dot system by using the density matrix formalism in Sect. 4.1. Here, to choose appropriate bases, which reflect symmetry of excited states, helps us to catch the physical meanings of the selective energy transfer, and also shows that the selective energy transfer can be controlled by adjusting spatial symmetry of quantum-dot arrangement. In Sect. 4.2, we provide concrete AND and XOR-logic operations in symmetrically arranged quantum-dot systems. Moreover, when the number of quantum dots, which form the coherent operation part increases, the degrees of freedom for selective energy transfer are extended. As an example, we numerically demonstrate controlled-type logic operations in Sect. 4.3, in which three identical quantum dots are utilized in the coherent operation part. Spatial symmetry is a key parameter in such a nanophotonic device using coherently coupled quantum dots because we can realize some strange operations mediating via so-called "dark state" [38]. Section 4.4 devotes to propose a nanophotonic buffer memory by using a spatially asymmetric system, which is an extremely interesting device because the device controls dissipation into a far-field photon reservoir. Furthermore, we show a device for identification of quantum entangled state in Sect. 4.5. Note that these logic and functional operations are in the irreversible process, although quantum entangled states are partially mediated to sort out information about initial excitations. This resembles quantum information processing, however, we do not require long coherence time as a

quantum computation. In regard to quantum information processing with dissipation or decoherence, there are several reports that are discussed tolerance and decoherence-free operations [39, 40].

4.1 Dynamics in a Coherently Coupled Quantum-Dot System

In order to examine, conditions of selective energy transfer from the coherent operation part to the dissipative output part, we derive analytic form of equations of motion in a simplest three-quantum-dot system, where two quantum dots are coherently coupled and a larger third quantum dots are assigned as the dissipative output part.

Symmetric and Anti-Symmetric States

Before a discussion about exciton population dynamics in a nanophotonic functional device using coherently coupled states, we explain several appropriate bases in a three-quantum-dot system, which reflect spatial symmetry among the three quantum dots. In such a system, we obtain a clear perspective of exciton dynamics by choosing the bases of the coupled states rather than those of isolated quantum-dot states which have been used in Sect. 3. From the symmetry of the system, the following bases are suitable for describing the dynamics of the one-exciton states using the smallest number of density matrix elements [41]:

$$|1\rangle_s = \frac{1}{\sqrt{2}} \left(|e\rangle_A |g\rangle_B |g, g\rangle_C + |g\rangle_A |e\rangle_B |g, g\rangle_C \right) , \tag{65}$$

$$|1\rangle_a = \frac{1}{\sqrt{2}} \left(|e\rangle_A |g\rangle_B |g, g\rangle_C - |g\rangle_A |e\rangle_B |g, g\rangle_C \right) , \tag{66}$$

$$|1\rangle_{ph} = |g\rangle_A |g\rangle_B |e, g\rangle_C , \tag{67}$$

$$|1\rangle_{pl} = |g\rangle_A |g\rangle_B |g, e\rangle_C , \tag{68}$$

where $|i, j\rangle_C$ $(i, j = g, e)$ represents the isolated quantum-dot states in QD-C with upper energy level i and lower energy level j. One-exciton state describes the condition whereby an exciton exists in either one of the three quantum dots. The crystal ground state and exciton state in each quantum dot, which is written as $|\Phi_g^\alpha\rangle$ and $|\Phi_{m(1s)}^\alpha\rangle$ in Sect. 2, are given by simplified form, such as $|g\rangle_A$ and $|e\rangle_A$, respectively. Similarly, a *two-exciton state* indicates that two excitons stay in the system. The suitable bases for the two-exciton states without occupation of the lower energy level in QD-C are expressed as

$$|2\rangle_{sh} = \frac{1}{\sqrt{2}} \left(|e\rangle_A |g\rangle_B |e, g\rangle_C + |g\rangle_A |e\rangle_B |e, g\rangle_C \right) , \tag{69}$$

$$|2\rangle_{ah} = \frac{1}{\sqrt{2}} \left(|e\rangle_A |g\rangle_B |e, g\rangle_C - |g\rangle_A |e\rangle_B |e, g\rangle_C \right) , \tag{70}$$

$$|2\rangle_{ph} = |e\rangle_A |e\rangle_B |g, g\rangle \tag{71}$$

and those with occupation of the lower energy level are expressed as

$$|2\rangle_{\text{sl}} = \frac{1}{\sqrt{2}}(|e\rangle_{\text{A}}|g\rangle_{\text{B}}|g, e\rangle_{\text{C}} + |g\rangle_{\text{A}}|e\rangle_{\text{B}}|g, e\rangle_{\text{C}}) , \tag{72}$$

$$|2\rangle_{\text{al}} = \frac{1}{\sqrt{2}}(|e\rangle_{\text{A}}|g\rangle_{\text{B}}|g, e\rangle_{\text{C}} - |g\rangle_{\text{A}}|e\rangle_{\text{B}}|g, e\rangle_{\text{C}}) , \tag{73}$$

$$|2\rangle_{\text{pl}} = |g\rangle_{\text{A}}|g\rangle_{\text{B}}|e, e\rangle_{\text{C}}, \tag{74}$$

where $|n\rangle_{\text{s}}$ and $|n\rangle_{\text{a}}$ ($n = 1, 2$) represent symmetric and anti-symmetric states in the coherent operation part, respectively, and the numbers 1 and 2 on the left-hand sides in (65–74) denote the one- and two-exciton states, respectively. In the following, first, we use these bases to investigate the exciton dynamics in a symmetrically arranged three-quantum-dot system. Then, we describe how to realize logic operation devices by using the symmetrically arranged quantum dots. Our discussion can expand into some interesting functional devices using spatial asymmetry of the system, which can be caught by the above characteristic bases.

Master Equation

By tracing out the photon degrees of freedom in the optical near-field coupling, which is formulated in Sect. 2, a model Hamiltonian for the three-quantum-dot system $\widehat{H}$ is given by

$$\widehat{H} = \widehat{H}_0 + \widehat{H}_{\text{int}} , \tag{75}$$

where

$$\widehat{H}_0 = \hbar\Omega_{\text{A}}\hat{A}^\dagger\hat{A} + \hbar\Omega_{\text{B}}\widehat{B}^\dagger\widehat{B} + \hbar\sum_{i=1}^{2}\Omega_{\text{C}_i}\hat{C}_i^\dagger\hat{C}_i , \tag{76}$$

$$\widehat{H}_{\text{int}} = \hbar U_{\text{AB}}(\hat{A}^\dagger\widehat{B} + \widehat{B}^\dagger\hat{A})$$
$$+\hbar U_{\text{BC}}(\widehat{B}^\dagger\hat{C}_2 + \hat{C}_2^\dagger\widehat{B}) + \hbar U_{\text{CA}}(\hat{C}_2^\dagger\hat{A} + \hat{A}^\dagger\hat{C}_2) , \tag{77}$$

where the definitions of the creation and annihilation operators, $(\hat{A}^\dagger, \hat{A})$, $(\widehat{B}^\dagger, \widehat{B})$, and $(\hat{C}_i^\dagger, \hat{C}_i)$, are shown schematically in Fig. 18. We assume that these are fermionic operators to give the effect of exciton–exciton interaction phenomenologically in the same energy level, but we neglect the exciton–exciton interaction between sublevels in QD-C. The eigenfrequencies for QD-A and B are represented by Ω_{A} and Ω_{B}, and the optical near-field couplings among three quantum dots are denoted as U_{AB}, U_{BC}, and U_{CA}. The equation of motion for the density operator of the quantum-dot system, $\hat{\rho}(t)$, is expressed by using the Born–Markov approximation [26] as

$$\dot{\hat{\rho}}(t) = -\frac{\text{i}}{\hbar}[\widehat{H}_0 + \widehat{H}_{\text{int}}, \hat{\rho}(t)]$$
$$+\frac{\Gamma}{2}\left\{2\hat{C}_1^\dagger\hat{C}_2\hat{\rho}(t)\hat{C}_2^\dagger\hat{C}_1 - \hat{C}_2^\dagger\hat{C}_1\hat{C}_1^\dagger\hat{C}_2\hat{\rho}(t) - \hat{\rho}(t)\hat{C}_2^\dagger\hat{C}_1\hat{C}_1^\dagger\hat{C}_2\right\} , \tag{78}$$

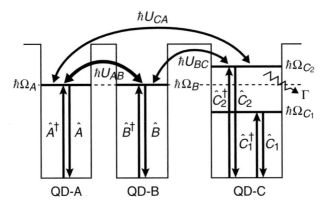

Fig. 18. Schematic drawing of exciton creation and annihilation operators and the energy transfer process in a three-quantum-dot system. The optical near-field coupling for the quantum-dot pairs are represented by U_{AB} for QD-A and B, U_{BC} for QD-B and C, and U_{CA} for QD-C and A. The nonradiative relaxation constant due to exciton–phonon coupling is denoted by Γ

where the nonradiative relaxation constant due to exciton–phonon coupling is denoted as Γ. The radiative relaxation due to exciton–photon coupling is omitted because the timescale of the optical near-field coupling and the exciton–phonon coupling is much faster than the radiative lifetime, which is in the order of a few nanoseconds. Taking matrix elements of (78) in terms of (65–68) after substituting (75–77) into (78), we obtain the following simultaneous differential equations for the one-exciton states:

$$\dot{\rho}_{s_1,s_1}(t) = i\sqrt{2}\overline{U}'(\rho_{s_1,ph_1}(t) - \rho_{ph_1,s_1}(t))$$
$$+i\Delta\Omega_{AB}(\rho_{s_1,a_1}(t) - \rho_{a_1,s_1}(t)) , \tag{79}$$

$$\dot{\rho}_{s_1,ph_1}(t) = \left\{ i(\Delta\Omega - U) - \frac{\Gamma}{2} \right\} \rho_{s_1,ph_1}(t)$$
$$+i\sqrt{2}\overline{U}'(\rho_{s_1,s_1}(t) - \rho_{ph_1,ph_1}(t))$$
$$-i\sqrt{2}\Delta U' \rho_{s_1,a_1}(t) - i\Delta\Omega_{AB}\rho_{a_1,p_1'}(t) , \tag{80}$$

$$\dot{\rho}_{ph_1,s_1}(t) = \left\{ -i(\Delta\Omega - U) - \frac{\Gamma}{2} \right\} \rho_{ph_1,s_1}(t)$$
$$-i\sqrt{2}\overline{U}'(\rho_{s_1,s_1}(t) - \rho_{ph_1,ph_1}(t))$$
$$+i\sqrt{2}\Delta U' \rho_{a_1,s_1}(t) + i\Delta\Omega_{AB}\rho_{p_1',a_1}(t) , \tag{81}$$

$$\dot{\rho}_{ph_1,ph_1}(t) = -\Gamma\rho_{ph_1,ph_1}(t) - i\sqrt{2}\overline{U}'(\rho_{s_1,ph_1}(t) - \rho_{ph_1,s_1}(t))$$
$$-i\Delta U'(\rho_{ph_1,a_1}(t) - \rho_{a_1,ph_1}(t)) , \tag{82}$$

$$\dot{\rho}_{a_1,ph_1}(t) = \left\{ i(\Delta\Omega + U) - \frac{\Gamma}{2} \right\} \rho_{a_1,ph_1}(t)$$
$$-i\sqrt{2}\Delta U'(\rho_{a_1,a_1}(t) - \rho_{ph_1,ph_1}(t))$$
$$+i\sqrt{2}\overline{U}' \rho_{a_1,s_1}(t) - i\Delta\Omega_{AB}\rho_{s_1,p'_1}(t) , \tag{83}$$

$$\dot{\rho}_{ph_1,a_1}(t) = \left\{ -i(\Delta\Omega + U) - \frac{\Gamma}{2} \right\} \rho_{ph_1,a_1}(t)$$
$$+i\sqrt{2}\Delta U'(\rho_{a_1,a_1}(t) - \rho_{ph_1,ph_1}(t))$$
$$-i\sqrt{2}\overline{U}' \rho_{s_1,a_1}(t) + i\Delta\Omega_{AB}\rho_{ph_1,s_1}(t) , \tag{84}$$

$$\dot{\rho}_{s_1,a_1}(t) = -i2U\rho_{s_1,a_1}(t) - i\sqrt{2}\overline{U}' \rho_{p'_1,a_1}(t) - i\sqrt{2}\Delta U'\rho_{s_1,ph_1}(t)$$
$$+i\Delta\Omega_{AB}(\rho_{s_1,s_1}(t) - \rho_{a_1,a_1}(t)) , \tag{85}$$

$$\dot{\rho}_{a_1,s_1}(t) = i2U\rho_{a_1,s_1}(t) + i\sqrt{2}\overline{U}' \rho_{a_1,ph_1}(t) + i\sqrt{2}\Delta U'\rho_{ph_1,s_1}(t)$$
$$-i\Delta\Omega_{AB}(\rho_{s_1,s_1}(t) - \rho_{a_1,a_1}(t)) , \tag{86}$$

$$\dot{\rho}_{a_1,a_1}(t) = i\sqrt{2}\Delta U'(\rho_{ph_1,a_1}(t) - \rho_{a_1,ph_1}(t))$$
$$-i\Delta\Omega_{AB}(\rho_{s_1,a_1}(t) - \rho_{a_1,s_1}(t)) , \tag{87}$$

where the density matrix element $_\alpha\langle n|\hat{\rho}(t)|n\rangle_\beta$ is abbreviated $\rho_{\alpha_n,\beta_n}(t)$ and the parameters, which have a dimension of frequency and characterize the dynamics in this system, are defined as $\Delta\Omega = \Omega_{C_2} - (\Omega_A + \Omega_B)/2$, $\Delta\Omega_{AB} = \Omega_A - \Omega_B$, $\overline{U}' = (U_{BC} + U_{CA})/2$, and $\Delta U' = (U_{BC} - U_{CA})/2$. The optical near-field coupling between QD-A and B, which is in a coherent operation part, is rewritten as $U = U_{AB}$. Similarly, in the case of two-exciton states described in (69–71), we obtain

$$\dot{\rho}_{sh_2,sh_2}(t) = -\Gamma\rho_{sh_2,sh_2}(t) + i\sqrt{2}\overline{U}'(\rho_{sh_2,ph_2}(t) - \rho_{ph_2,sh_2}(t))$$
$$+i\Delta\Omega_{AB}(\rho_{sh_2,ah_2}(t) - \rho_{ah_2,sh_2}(t)) , \tag{88}$$

$$\dot{\rho}_{sh_2,ph_2}(t) = \left\{ -i(\Delta\Omega + U) - \frac{\Gamma}{2} \right\} \rho_{sh_2,ph_2}(t)$$
$$+i\sqrt{2}\overline{U}'(\rho_{sh_2,sh_2}(t) - \rho_{ph_2,ph_2}(t))$$
$$+i\sqrt{2}\Delta U'\rho_{sh_2,ah_2}(t) - i\Delta\Omega_{AB}\rho_{ah_2,ph_2}(t) , \tag{89}$$

$$\dot{\rho}_{ph_2,sh_2}(t) = \left\{ i(\Delta\Omega + U) - \frac{\Gamma}{2} \right\} \rho_{ph_2,sh_2}(t)$$
$$-i\sqrt{2}\overline{U}'(\rho_{sh_2,sh_2}(t) - \rho_{ph_2,ph_2}(t))$$
$$-i\sqrt{2}\Delta U'\rho_{ah_2,sh_2}(t) + i\Delta\Omega_{AB}\rho_{ph_2,ah_2}(t) , \tag{90}$$

$$\dot{\rho}_{\mathrm{ph}_2,\mathrm{ph}_2}(t) = -\mathrm{i}\sqrt{2}\overline{U}'(\rho_{\mathrm{sh}_2,\mathrm{ph}_2}(t) - \rho_{\mathrm{ph}_2,\mathrm{sh}_2}(t))$$

$$+\mathrm{i}\sqrt{2}\Delta U'(\rho_{\mathrm{ph}_2,\mathrm{ah}_2}(t) - \rho_{\mathrm{ah}_2,\mathrm{ph}_2}(t)) , \tag{91}$$

$$\dot{\rho}_{\mathrm{ah}_2,\mathrm{ph}_2}(t) = \left\{-\mathrm{i}(\Delta\Omega - U) - \frac{\Gamma}{2}\right\}\rho_{\mathrm{ah}_2,\mathrm{ph}_2}(t)$$

$$+\mathrm{i}\sqrt{2}\Delta U'(\rho_{\mathrm{ah}_2,\mathrm{ah}_2}(t) - \rho_{\mathrm{ph}_2,\mathrm{ph}_2}(t))$$

$$+\mathrm{i}\sqrt{2}\overline{U}'\rho_{\mathrm{ah}_2,\mathrm{sh}_2}(t) - \mathrm{i}\Delta\Omega_{\mathrm{AB}}\rho_{\mathrm{sh}_2,\mathrm{ph}_2}(t) , \tag{92}$$

$$\dot{\rho}_{\mathrm{ph}_2,\mathrm{ah}_2}(t) = \left\{\mathrm{i}(\Delta\Omega - U) - \frac{\Gamma}{2}\right\}\rho_{\mathrm{ph}_2,\mathrm{ah}_2}(t)$$

$$-\mathrm{i}\sqrt{2}\Delta U'(\rho_{\mathrm{ah}_2,\mathrm{ah}_2}(t) - \rho_{\mathrm{ph}_2,\mathrm{ph}_2}(t))$$

$$-\mathrm{i}\sqrt{2}\overline{U}'\rho_{\mathrm{sh}_2,\mathrm{ah}_2}(t) + \mathrm{i}\Delta\Omega_{\mathrm{AB}}\rho_{\mathrm{ph}_2,\mathrm{sh}_2}(t) , \tag{93}$$

$$\dot{\rho}_{\mathrm{sh}_2,\mathrm{ah}_2}(t) = (-\mathrm{i}2U - \Gamma)\rho_{\mathrm{sh}_2,\mathrm{ah}_2}(t)$$

$$-\mathrm{i}\sqrt{2}\overline{U}'\rho_{\mathrm{ph}_2,\mathrm{ah}_2}(t) + \mathrm{i}\sqrt{2}\Delta U'\rho_{\mathrm{sh}_2,\mathrm{ph}_2}(t)$$

$$+\mathrm{i}\Delta\Omega_{\mathrm{AB}}(\rho_{\mathrm{sh}_2,\mathrm{sh}_2}(t) - \rho_{\mathrm{ah}_2,\mathrm{ah}_2}(t)) , \tag{94}$$

$$\dot{\rho}_{\mathrm{ah}_2,\mathrm{sh}_2}(t) = (\mathrm{i}2U - \Gamma)\rho_{\mathrm{ah}_2,\mathrm{sh}_2}(t)$$

$$+\mathrm{i}\sqrt{2}\overline{U}'\rho_{\mathrm{ah}_2,\mathrm{ph}_2}(t) - \mathrm{i}\sqrt{2}\Delta U'\rho_{\mathrm{ph}_2,\mathrm{sh}_2}(t)$$

$$-\mathrm{i}\Delta\Omega_{\mathrm{AB}}(\rho_{\mathrm{sh}_2,\mathrm{sh}_2}(t) - \rho_{\mathrm{ah}_2,\mathrm{ah}_2}(t)) , \tag{95}$$

$$\dot{\rho}_{\mathrm{ah}_2,\mathrm{ah}_2}(t) = -\Gamma\rho_{\mathrm{ah}_2,\mathrm{ah}_2}(t) - \mathrm{i}\sqrt{2}\Delta U'(\rho_{\mathrm{ph}_2,\mathrm{ah}_2}(t) - \rho_{\mathrm{ah}_2,\mathrm{ph}_2}(t))$$

$$-\mathrm{i}\Delta\Omega_{\mathrm{AB}}(\rho_{\mathrm{sh}_2,\mathrm{ah}_2}(t) - \rho_{\mathrm{ah}_2,\mathrm{sh}_2}(t)) . \tag{96}$$

From (79–96), we can attract three important features with regard to energy transfer from the coherent operation part to the dissipative output part. Comparing (80) with (83), where the diagonal elements represent the transition probability, opposite contributions appear in the first terms, i.e., $\Delta\Omega - U$ and $\Delta\Omega + U$. These correspond to the difference of resonance conditions for symmetric and anti-symmetric states, respectively. The resonance conditions are easily interpreted by considering the energies for the symmetric and anti-symmetric states, which can be derived from (75–77) as

$$\langle s_1|\widehat{H}|s_1\rangle = \hbar(\bar{\Omega}_{\mathrm{AB}} + U) , \tag{97}$$

$$\langle a_1|\widehat{H}|a_1\rangle = \hbar(\bar{\Omega}_{\mathrm{AB}} - U) , \tag{98}$$

where $\bar{\Omega}_{\mathrm{AB}} = (\Omega_{\mathrm{A}} + \Omega_{\mathrm{B}})/2$. Figure 19a is the schematic illustration that explains relation between isolated quantum-dot states and coherently coupled

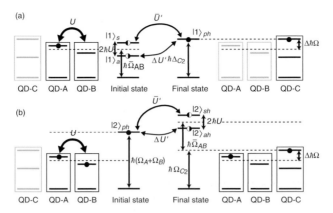

Fig. 19. Schematic explanation of a relation between isolated quantum-dot states and coupled states for **(a)** one- and **(b)** two-exciton states. The *left* and *right* illustrations represent the initial and final states, respectively, and the center figures denote the coupled states for input and output. The energy transfer between symmetric states $|n\rangle_s$ and output states $|n\rangle_{ph}$, where $n = 1, 2$, occurs via coupling strength $\hbar U'$, and that between asymmetric states $|n\rangle_a$ and output states $|n\rangle_{ph}$ does via $\hbar \Delta U'$

states. A filled circle indicates that an exciton occupies the corresponding energy level, while a semicircle indicates that an exciton exists in the energy level in either QD-A or B with a certain probability. The first feature is that we can select the symmetric and anti-symmetric states by adjusting the energy level in QD-C, where the energy shift $\Delta \Omega$ is determined by the strength of the optical near-field coupling U.

The second feature is also observed from the second terms in (80) and (83). These terms determine the strength or the speed of resonant energy transfer between the coherent operation part and dissipative output part. The difference is as follows: the symmetric state resonantly couples to the upper energy level in QD-C, i.e., $|1\rangle_{ph}$-state, by mediating the averaged coupling strength $\overline{U}'$, while the anti-symmetric state does by mediating the difference between two coupling paths, $\Delta U'$. When the three quantum dots are arranged symmetrically in nanometric space, the energy transfer between $|1\rangle_a$-state and $|1\rangle_{ph}$-state is forbidden because of $\Delta U' = 0$, in other words, $_{ph}\langle 1|\hat{H}_{int}|1\rangle_a = 0$. Therefore, we can control the energy transfer or signal flow by using spatial symmetry in the system.

The third feature is related to the exciton numbers in the system, which is interpreted by comparing (80) and (89) or (83) and (92). Similar to the first feature, the resonance conditions for the energy transfer from the coherent operation part to the dissipative output part indicate opposite contribution, i.e., $\Delta \Omega - U$ and $\Delta \Omega + U$ for one- and two-exciton states, respectively. In Fig. 19b, the correspondence of isolated quantum-dot states and coupled states

are schematically drawn, where we can see the inversion of the resonance conditions. The difference of the resonance conditions is applicable to two input logic operations discussed later. Note that the first two features, which are discussed for only one-exciton state, are reversed in the case of two-exciton state as you observed in (88–96).

4.2 Nanophotonic Logic Gates

In order to realize well-known AND- and XOR-logic operations, we assume spatially symmetric quantum-dot system, and thus, the related differential equations in (79–96) are restricted, which decouples the anti-symmetric $|1\rangle_a$- and $|2\rangle_a$-states from the above. In this case, the dynamics can be solved analytically with the help of Laplace transforms for typical initial conditions. The output population for the one-exciton state can be written as

$$
\begin{aligned}
\rho_{\mathrm{pl}_1,\mathrm{pl}_1}(t) &= \Gamma \int_0^t \rho_{\mathrm{ph}_1,\mathrm{ph}_1}(t')\mathrm{d}t' \\
&= \frac{1}{2} + \frac{4U'^2}{\omega_+^2 - \omega_-^2}\mathrm{e}^{-(\Gamma/2)t} \\
&\quad \times \{\cos\phi_+ \cos(\omega_+ t + \phi_+) - \cos\phi_- \cos(\omega_- t + \phi_-)\}
\end{aligned} \tag{99}
$$

with

$$
\begin{aligned}
\omega_\pm &= \frac{1}{\sqrt{2}}\left[(\Delta\Omega - U)^2 + W_+ W_- \right. \\
&\quad \left. \pm\sqrt{\{(\Delta\Omega - U)^2 + W_+^2\}\{(\Delta\Omega - U)^2 + W_-^2\}}\right]^{1/2}, \\
\phi_\pm &= \tan^{-1}\left(\frac{2\omega_\pm}{\Gamma}\right), \\
W_\pm &= 2\sqrt{2}U' \pm \frac{\Gamma}{2}
\end{aligned} \tag{100}
$$

for the initial condition $\rho_{\mathrm{s}_1,\mathrm{s}_1}(0) = \rho_{\mathrm{a}_1,\mathrm{a}_1}(0) = \rho_{\mathrm{s}_1,\mathrm{a}_1}(0) = \rho_{\mathrm{a}_1,\mathrm{s}_1}(0) = 1/2$, which corresponds to the condition $_A\langle e|_B\langle g|_C\langle g, g|\hat{\rho}(t)|e\rangle_A|g\rangle_B|g, g\rangle_C = 1$ and otherwise zero. The notation of optical near-field coupling is rewritten as $U' = \overline{U}'$ because the coupling strengths between QD-B and C, and QD-C and A are equivalent for symmetrically arranged system, that is $U' = U_{\mathrm{BC}} = U_{\mathrm{CA}}$. The first line in (99) denotes the irreversible process of nonradiative relaxation, which is easily interpreted from the temporal sequence of the one-exciton state illustrated in Fig. 20a.

Analytic solutions for two-exciton states can be obtained from an equation similar to (99), except for the sign of U, i.e., with the resonance conditions inverted. The probability of an exciton occupying the lower energy level

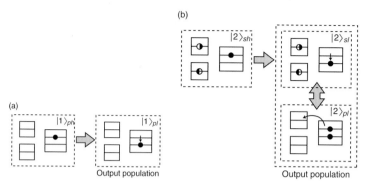

Fig. 20. Schematic illustration of temporal sequences at the final output stages in the cases of (**a**) one-exciton state and (**b**) two-exciton state

in QD-C is

$$\rho_{sl_2,sl_2}(t) + \rho_{pl_2,pl_2}(t)\Gamma \int_0^t \rho_{sh_2,sh_2}(t')dt'$$

$$= 2\left[\frac{1}{2} + \frac{4U'^2}{\omega'_+{}^2 - \omega'_-{}^2}e^{-(\Gamma/2)t}\right.$$

$$\times\left\{\cos\phi'_+\cos(\omega'_+t+\phi'_+) - \cos\phi'_-\cos(\omega'_-t+\phi'_-)\right\}\right] \qquad (101)$$

with

$$\omega'_\pm = \frac{1}{\sqrt{2}}\left[(\Delta\Omega + U)^2 + W_+W_-\right.$$

$$\left.\pm\sqrt{\{(\Delta\Omega + U)^2 + W_+^2\}\{(\Delta\Omega + U)^2 + W_-^2\}}\right]^{1/2},$$

$$\phi'_\pm = \tan^{-1}\left(\frac{2\omega'_\pm}{\Gamma}\right), \qquad (102)$$

where the factor 2 in (101) comes from the initial conditions for the two-exciton state, i.e., $\rho_{ph_2,ph_2}(0) = 1$ and otherwise zero. As you can see from the right hand side in the first line in (101), there are two final output states, $|2\rangle_{sl}$ and $|2\rangle_{pl}$. However, the previous excited state, $|2\rangle_{sh}$, only contributes to the output population, because the state $|2\rangle_{pl}$ is made from $|2\rangle_{sl}$, and thus, the total population does not change, which is illustrated in Fig. 20b. In both (99) and (101), the second terms with the denominators $\omega_+^2 - \omega_-^2$ and $\omega'_+{}^2 - \omega'_-{}^2$, respectively, contribute to the increase of population in the output energy levels. When we set $\Delta\Omega = U$, efficiency of energy transfer is dominant for the one-exciton state because $\omega_+^2 - \omega_-^2$ becomes minimum, while that is dominant for the two-exciton state in the case of $\Delta\Omega = -U$. These are analytical description of resonant energy transfer depending on the number of

input excitons and a basic idea for logic operation. In the following, we explain AND- and XOR-logic operations schematically, and show concrete calculated results of them.

AND-Logic Operation

When the upper energy level in QD-C is negatively shifted, which corresponds to $\hbar(\Omega - U)$, we can realize an AND-logic gate. Figure 21 represents energy diagram in the system with negative detuning in QD-C. As you can see from Fig. 21a, the resonant energy transfer occurs only for the two-exciton state via the symmetric state in QD-A and QD-B, while the energy transfer for the one-exciton state does not because the energy level in QD-C is resonant to the anti-symmetric state which is dipole inactive state for the symmetrically arranged quantum-dot system. This characteristic selective energy transfer assures an AND-logic operation.

The temporal evolution of the exciton population on the lower energy level in QD-C, which is analytically derived in (99) and (101), is plotted in Fig. 22, where the strengths of optical near-field coupling, $\hbar U = 89\,\mu\text{eV}$ and $\hbar U' = 14\,\mu\text{eV}$, referred to the estimated values in Sect. 2 for CuCl quantum cubes embedded in NaCl matrix. As we discussed above, the coupling to far-field light is neglected because our interests are fast population dynamics due to the optical near-field coupling, which is in the order of sub-100 ps. In Fig. 22, the exciton population is almost occupied at about 100 ps for the two-exciton state, which is determined by the coupling strength $\hbar U'$ between QD-A and C (QD-B and C). The negative energy shift is also set as $89\,\mu\text{eV}$, and the nonradiative relaxation is 10 ps.

From Fig. 22, the output population can be observed only for the two-exciton state as we expected, because the coupling occurs via the symmetric

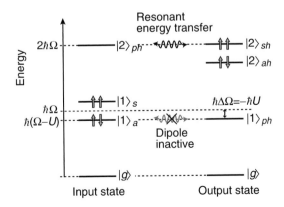

Fig. 21. Energy diagram in a coupled three-quantum-dot system with negative energy detuning in QD-C

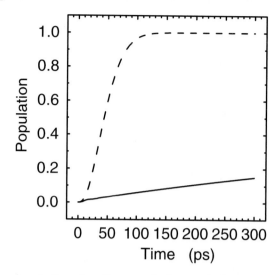

Fig. 22. Temporal evolution of exciton population on the lower energy level (output level) in QD-C. The *solid* and *dashed curves* represent the cases for one- and two-exciton states, respectively. The coupling strength between QD-A and B is set as $\hbar U = 89\,\mu\text{eV}$, and that between QD-B and C (QD-A and C) is $U' = 14\,\mu\text{eV}$

state. On the other hand, in the case of the one-exciton state, the population increases very slowly. This is caused by the weak coupling between the symmetric state in the input side and the output state. The state-filling time is much longer than spontaneous emission lifetime, and thus, the exciton population for the one-exciton state does not affect the output signal, that is OFF-level. In this manner, these operations for the one- and two-exciton states surely correspond to the AND-logic gate whose size is much smaller than light diffraction limits.

XOR-Logic Operation

Opposite to the AND-logic gate, suppose a system in which the upper energy level in QD-C is positively shifted, i.e., $\hbar(\Omega + U)$. Energy diagram in this system is illustrated in Fig. 23. In this case, the symmetric and anti-symmetric states for one and two excitons satisfy the conditions for an XOR-logic gate. The energy transfer from the input system to the output system allows when an exciton is excited in either QD-A or QD-B, while, for the two-exciton state, the anti-symmetric state in the output system is dipole inactive against the input state in the symmetrically arranged quantum-dot system.

Figure 24 shows analytic curves of the temporal evolution for the one- and two-exciton states. The given parameters, such as the coupling strength via the optical near field, nonradiative relaxation time, are completely the same as those in Fig. 22. In Fig. 24, the output population appears for the

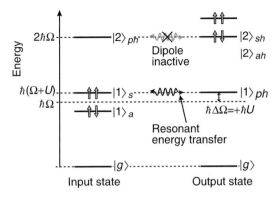

Fig. 23. Energy diagram in a coupled three quantum-dot system with positive energy detuning in QD-C

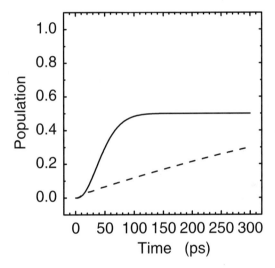

Fig. 24. Temporal evolution of exciton population on the lower energy level (output level) in QD-C. The *solid* and *dashed curves* represent the cases for one- and two-exciton states, respectively. The coupling strength between QD-A and B is set as $U = 89\,\mu\text{eV}$, and that between QD-B and C (QD-A and C) is $U' = 14\,\mu\text{eV}$

one-exciton state, and the state-filling time is determined by the coupling strength U' between QD-A and C (QD-B and C), which is the same as the AND-logic gate. The readers immediately put the question why the exciton population reaches a half of the maximum for the one-exciton state in the XOR-logic gate. This is because a one-side quantum dot is locally excited as an initial condition, in which both the symmetric and anti-symmetric states are simultaneously excited with the same occupation probability as described in (65) and (66). On the other hand, we can observe slow increase of exciton population for the two-exciton state (OFF-level) in Fig. 22, which is as twice fast as that for the AND-logic gate. This also originates from the initial excited state; the symmetric state in the input system is occupied the half of the maximum at the initial time because of local excitation, while the full of the initial population for the two-exciton state can couple to QD-C in the case of positive energy shift.

The above behavior that the output population detected when a one-side quantum dot in the input system is initially excited corresponds to an XOR-logic gate as a stochastic meaning. This means that the fully occupied output state cannot achieve for a single exciton process.

Here, we summarize the operation of our proposed AND- and XOR-logic gates in Table 1, which are inherent operations in nanophotonic devices using typical coherent and dissipative process. The system behaves as an AND-logic gate when the energy difference is set to $\Delta\Omega = -U$, and the system provides an XOR-like-logic operation when $\Delta\Omega = U$. It is noteworthy that these operations are different from the quantum logic operation, because long quantum coherence time is unnecessary. The critical limit of these logic gates is determined by the following condition; the energy transfer time from the coherent operation part to the dissipative output part, which is estimated about 50 ps for the CuCl quantum-cube system, is enough shorter than the radiative lifetime ($\sim$1 ns) of excitons in each quantum dot.

Signal Contrast

The steepness of the resonance determines the contrast of the output signal. In order to discuss how to obtain high contrast signal, dependence of the exciton population on the energy shift $\Delta\Omega$ in the above symmetrically arranged quantum-dot system is plotted in Fig. 25. The longitudinal axis is the population at $t = 100$ ps, which is the time until energy transfer almost finishes under resonance conditions $\Delta\Omega = \pm U$ and is analytically derived from (99) and (101). We clearly find that two types of switching operations, i.e., AND- and XOR-logic operations, can be realized by choosing the energy shift as $\Delta\Omega = \pm U$. Here, the peak width of both curves in Fig. 25 is given by the product $W_+ W_-$ in (100) or (102), which corresponds to a balance between

Table 1. Relationship between the input and output populations for the energy difference $\Delta\Omega = \pm U$

input		output: C	
A	B	$\Delta\Omega = -U$	$\Delta\Omega = U$
0	0	0	0
1	0	0	0.5
0	1	0	0.5
1	1	1	0

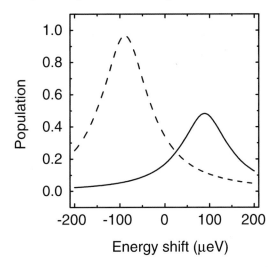

Fig. 25. Variation in the output populations at a fixed time of $t = 100\,\mathrm{ps}$ as a function of the energy shift $\Delta\Omega$. The *solid* and *dashed curves* represent the one- and two-exciton states, respectively. The optical near-field coupling strengths $\hbar U$ and $\hbar U'$, and the nonradiative relaxation constant Γ have the same values as in Fig. 22

the coupling strength U' between the coherent operation part and the dissipative output part and the nonradiative relaxation constant Γ. Therefore, narrow peaks are obtained when the conditions $W_+ \ll 1$ and $W_- = 0$, i.e., $2\sqrt{2}U' \sim \Gamma/2 \ll 1$, are satisfied. In this case, the highest contrast of the logic operations can be achieved.

Effects of Asymmetry

It is valuable to examine the exciton dynamics in an asymmetrically arranged quantum-dot system to estimate the fabrication tolerance for the system described above and to propose further functional operations inherent to nanophotonic devices which we will discuss in Sects. 4.4 and 4.5. Here, we investigate the effects of asymmetry numerically. In addition, we comment on a positive use of these effects. When the three quantum dots are arranged asymmetrically, we must consider the dynamics of all density matrix elements given in (79–96), because the exciton population leaks to the anti-symmetric states, which decouples in the case of symmetrically arranged system. In the asymmetrically arranged system, the exciton dynamics between states $|n\rangle_\mathrm{s}$ and $|n\rangle_\mathrm{ph}$ do not change from the symmetrically arranged system, where the coupling strength is replaced by the average value $\overline{U}'$. The main difference is that the matrix elements for states $|n\rangle_\mathrm{a}$ can couple with states $|n\rangle_\mathrm{s}$ and $|n\rangle_\mathrm{ph}$ in the asymmetrically arranged system, while these are decoupled in the symmetrically arranged system. Two types of coupling emerge in the asymmetric system: one originates from the energy difference $\Delta\Omega_{AB}$ between QD-A and B, and the other comes from the arrangement of the three quantum dots, which is expressed using the parameter $\Delta U'$. Previously [7], we discussed the influence of the energy difference on the exciton dynamics in a two-quantum-dot system that mainly degrades the signal contrast. Here, we focus on the effects of asymmetry due to the spatial arrangement of each quantum dot, by assuming $\Delta\Omega_{AB} = 0$.

In order to examine the effects of the quantum-dot arrangement, the average coupling strength $\overline{U}'$ is fixed so that states $|n\rangle_s$ and $|n\rangle_{ph}$ maintain the same temporal evolution that was found in the symmetrically arranged system. Then, the difference between the coupling strengths $\Delta U'$ varies from 0 to $\pm U$, where the exciton dynamics are independent of the sign of $\Delta U'$. Therefore, an asymmetry factor is defined by the ratio of $|\Delta U'|$ to $\overline{U}'$, varying from 0 (symmetry) to 1 (maximum amount of asymmetry).

Figure 26 presents the temporal evolution of the output population for the energy shift $\Delta\Omega = -U$ (an AND-logic gate case) with and without an asymmetric arrangement. For the one-exciton state (Fig. 26a), the asymmetric arrangement strongly affects the exciton dynamics, and the OFF-state in the AND-logic gate operation is no longer valid because the off-resonance condition for the energy transfer between states $|1\rangle_s$ and $|1\rangle_{ph}$ acts oppositely to the resonance condition between states $|1\rangle_a$ and $|1\rangle_{ph}$. This is evident in (79–87), for example, by comparing the matrix elements $\rho_{s_1,ph_1}(t)$ with $\rho_{a_1,ph_1}(t)$. Therefore, in the one-exciton state, the exciton population is very sensitive to the asymmetric arrangement. By contrast, the two-exciton state

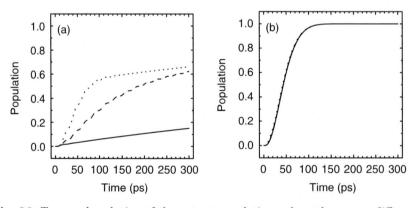

Fig. 26. Temporal evolution of the output populations where the energy difference is set to $\Delta\Omega = -U$ ($\hbar U = 89\,\mu eV$). **(a)** and **(b)** show the populations for one- and two-exciton states, respectively. The *solid, dashed,* and *dotted curves* represent the results for asymmetry factors $\Delta U'/\overline{U}' = 0$, 0.5, and 1.0, respectively, where the average coupling strength is set to $\hbar\overline{U}' = 14\,\mu eV$. In **(b)**, the three curves are almost identical

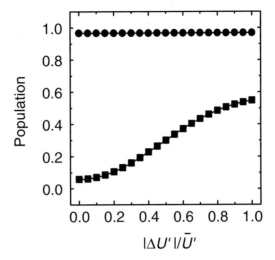

Fig. 27. Variation in the output populations at a fixed time of $t = 100\,\mathrm{ps}$ as a function of the asymmetry factor, where the energy difference is set to $\Delta\Omega = -U$ ($\hbar U = 89\,\mu\mathrm{eV}$) and an average coupling strength of $\hbar\bar{U} = 14\,\mu\mathrm{eV}$ is used. The *curves* shown with *square* and *circular dots* represent the one- and two-exciton states, respectively. Only the exciton population in the one-exciton state is modified by increasing the asymmetry factor

is not influenced by the quantum-dot arrangement (see Fig. 26b). We also observe small and high-frequency oscillations for the dashed and dotted curves ($|\Delta U'|/\bar{U}' = 0.5$ and 1.0) in Fig. 26a. These come from the coherence between states $|1\rangle_\mathrm{s}$ and $|1\rangle_\mathrm{a}$ which can be seen in the equations of motion of $\rho_{s_1,a_1}(t)$ and $\rho_{a_1,s_1}(t)$. Since the coherence is always excited by mediating state $|1\rangle_\mathrm{ph}$, and the state $|1\rangle_\mathrm{ph}$ has a short lifetime dominated by the relaxation constant Γ, the oscillations have no relation to the population dynamics. Figure 27 shows the variation in the output population at $t = 100\,\mathrm{ps}$ as a function of the asymmetry factor $|\Delta U'|/\bar{U}'$. From this figure, it follows that the asymmetry only affects the one-exciton state, where it breaks the OFF-state in the logic gate, as shown by the curve with squares, and the signal contrast decreases with increasing asymmetry.

Conversely, for the XOR-logic gate ($\Delta\Omega = U$), the two-exciton states correspond to the off-resonant states in the symmetric system. Therefore, the excitation is transferred to the output energy level in QD-C as the asymmetry factor increases, as shown in Fig. 28b. Similarly, the variation in the output population with the asymmetry factor is plotted in Fig. 29, where the time is fixed at $t = 100\,\mathrm{ps}$. The figure shows that the XOR-logic operation in the symmetric system is reversed when the asymmetry factor exceeds 0.5 because a one-exciton state can occupy the initial state of $|1\rangle_\mathrm{s}$ with a probability of $1/2$, as shown above. Consequently, the output population also reaches a

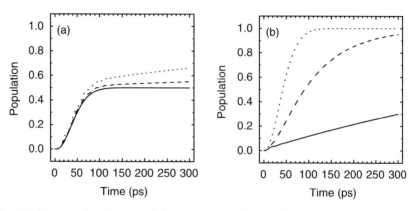

Fig. 28. Temporal evolution of the output populations for an the energy difference of $\Delta\Omega = U$ ($\hbar U = 89\,\mu\text{eV}$). **(a)** and **(b)** show the populations for the one- and two-exciton states, respectively. The *solid*, *dashed*, and *dotted curves* represent the results for asymmetry factors $\Delta U'/\overline{U}' = 0$, 0.5, and 1.0, respectively, where the average coupling strength is set to $\hbar\overline{U}' = 14\,\mu\text{eV}$

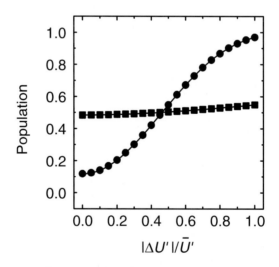

Fig. 29. Variation in the output populations at the fixed time of $t = 100\,\text{ps}$ as a function of the asymmetry factor, where the energy difference is set to $\Delta\Omega = U$ ($\hbar U = 89\,\mu\text{eV}$) and an average coupling strength of $\hbar\overline{U} = 14\,\mu\text{eV}$ is used. The curves shown with *square* and *circular dots* represent the one- and two-exciton states, respectively. The exciton population in the two-exciton state exceeds that of the one-exciton state when the asymmetry factor $\Delta U'/\overline{U}'$ exceeds 0.5, so the XOR-logic operation is reversed

probability of 1/2. This is also valid in the asymmetric system. However, the asymmetric arrangement enables coupling of the two-exciton states, $|2\rangle_{\text{ph}}$ and $|1\rangle_{\text{ah}}$. State $|2\rangle_{\text{ph}}$ can be fully excited in the initial stage, so the output population reaches a unit value via states $|2\rangle_{\text{ah}}$. This exceeds the output population 0.5 for a one-exciton state with a lager amount of asymmetry.

Although spatial asymmetry acts negatively for the above logic operations by using a symmetrically arranged system, while this has hidden potential toward nanophotonic inherent functions. As mentioned above, the effect of asymmetry is based on coupling to states $|n\rangle_{\text{a}}$ in an asymmetrically arranged system, which are so-called "dark states" [38]. If we create such dark states by using optical near-field interaction just as we intended, which cannot be excited by far-field light, confinement of photons in a nanophotonic device can be realized, which is discussed in the following. Furthermore, in such an asymmetrically arranged system with via coherent excitations, the symmetric and anti-symmetric states can be excited partially, where both states are expressed by the superposition of eigenstates in isolated (noninteracting) quantum dots. Therefore, a system composed of three quantum dots can not only select information that depends on the initially prepared excitations, but also information that reflects the initial quantum entangled states in the coherent operation part. From this perspective, such nanophotonic devices are useful in connecting quantum devices as a detector and interface devices which identify occupation probability of the quantum entangled states in an input signal.

4.3 Nanophotonic Controlled Logic Gates

Up to this point, we have instructively discussed the simplest system with three quantum dots. Focused on a coherent operation part, the readers easily understand that more coupled states can be prepared when the number of quantum dots increases; for example, the coherent operation part which consists of three identical quantum dots, i.e., a four-quantum-dot system illustrated in Fig. 30a, has three coupled states

$$|1\rangle_{\text{su}} = \frac{1}{2}(|e\rangle_A|g\rangle_B|g\rangle_C + \sqrt{2}|g\rangle_A|e\rangle_B|g\rangle_C + |g\rangle_A|g\rangle_B|e\rangle_C) , \qquad (103)$$

$$|1\rangle_{\text{a}} = \frac{1}{\sqrt{2}}(|e\rangle_A|g\rangle_B|g\rangle_C - |g\rangle_A|g\rangle_B|e\rangle_C) , \qquad (104)$$

$$|1\rangle_{\text{sl}} = \frac{1}{2}(|e\rangle_A|g\rangle_B|g\rangle_C - \sqrt{2}|g\rangle_A|e\rangle_B|g\rangle_C + |g\rangle_A|g\rangle_B|e\rangle_C) \qquad (105)$$

for the one-exciton states, and

$$|2\rangle_{\text{su}} = \frac{1}{2}(|e\rangle_A|e\rangle_B|g\rangle_C + \sqrt{2}|e\rangle_A|g\rangle_B|e\rangle_C + |g\rangle_A|e\rangle_B|e\rangle_C) , \qquad (106)$$

$$|2\rangle_{\text{a}} = \frac{1}{\sqrt{2}}(|e\rangle_A|e\rangle_B|g\rangle_C - |g\rangle_A|e\rangle_B|e\rangle_C) , \qquad (107)$$

$$|2\rangle_{\text{sl}} = \frac{1}{2}(|e\rangle_A|e\rangle_B|g\rangle_C - \sqrt{2}|e\rangle_A|g\rangle_B|e\rangle_C + |g\rangle_A|e\rangle_B|e\rangle_C) \qquad (108)$$

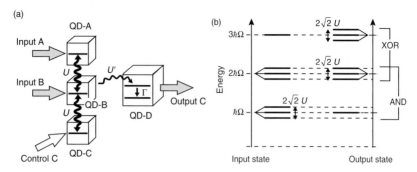

Fig. 30. (a) Schematic illustration of four-quantum-dot system, which consists of three identical quantum dots as the coherent operation part and a quantum dot with energy sublevels as the dissipative output part. (b) Energy diagram in the four-quantum-dots system. The *left* and *right* diagrams correspond to the input and output states, respectively, and the energy are split into three levels with the energy shift of $\sqrt{2}U$ for the one- and two-exciton states in the input state and for the two- and three-exciton states in the output state

for the two-exciton states. A three-exciton state is degenerated since the energy levels in all three quantum dots are occupied. By using these coupled states, we can propose a characteristic functional device, which is named controlled AND- and XOR-logic gates. Although equations of motion for density matrix elements build up in a similar manner to (79–96), it is tedious to describe differential equations for all density matrix elements. Here we explain energy transfer property by using schematic illustration. The energy diagram is illustrated in Fig. 30b, where the energies for one- and two-exciton states are divided into three levels with eigenenergies, $\hbar\Omega + \sqrt{2}$, $\hbar\Omega$, and $\hbar\Omega - \sqrt{2}$ due to the optical near-field coupling. Note that the coupled state with middle energy is dipole inactive state in a symmetric arranged system (Fig. 30a). The left side in the diagram corresponds to the input (initial) state, where only the 'three-exciton state is degenerated, while the right denotes the output state in which the three-exciton state becomes nondegenerated since an exciton stays in QD-D without going back to the coherent operation part because of fast intra-sublevel relaxation.

In Fig. 31, we show the result of output population at 300 ps as a function of the energy shift $\Delta\Omega$, where initial states are set as $(1,0,0)$, $(1,1,0)$, and $(1,1,1)$, the notation corresponding to the quantum-dot label of (A, B, C). In order to obtain clear energy splitting (sharp resonance) for the one-, two-, and three-exciton states, the strength of optical near-field coupling is set as a suitably optimized value of $\hbar U = 66\,\mu\mathrm{eV}$ (10 ps). This seems somewhat strong as compared with the previously estimated value for CuCl quantum cubes embedded in an NaCl matrix. Therefore, we require some optimization in materials and size of quantum dots to realize this controlled-logic devices. From Fig. 31, we understand that only two-exciton states can coupled to the output state, when we choose the upper energy level in QD-D equivalent to

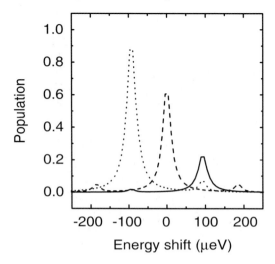

Fig. 31. Variation in the output populations at a fixed time of $t = 300\,\mathrm{ps}$ as a function of the energy difference $\Delta\Omega$. The *solid, dashed,* and *dotted curves* represent the one-, two-, and three-exciton states, respectively. The optical near-field coupling strengths $\hbar U$ and $\hbar U'$, and the nonradiative relaxation constant Γ are set as 66, $3.3\,\mu\mathrm{eV}$, and $(50\,\mathrm{ps})^{-1}$, respectively

the middle energy level in the coupled states, no energy shift being applied. Regarding to the input terminals as QD-A and B, the exciton population transfers to the output terminal of QD-D differently, whether the control terminal of QD-C is excited or not. In other words, QD-C plays a role to exchange the AND- and XOR-logic operations.

In the following, we discussed temporal evolution of exciton population on the output energy level in QD-D, which is numerically derived by using bases of isolated quantum dots. Figure 32 shows the results of calculation for all possible initial excitations in the symmetrically arranged four-quantum-dot system. The parameters are set as the same values in Fig. 31. At first, we focus on the exciton dynamics when the population in QD-C (control dot) is empty at the initial time. In this case, an exciton which is prepared in QD-A or QD-B cannot move to the output energy level in QD-D. The reason is as follows: a locally excited state in QD-A is expressed as a superposition of the coupled states, $|e\rangle_A|g\rangle_B|g\rangle_C = (|1\rangle_{su} + \sqrt{2}|1\rangle_a + |1\rangle_{sl})/2$, where the output energy level is resonant for the state $|1\rangle_a$ and off-resonant for the states $|1\rangle_{su}$ and $|1\rangle_{sl}$ as you can see in Fig. 30. However, the state $|1\rangle_a$ is dipole inactive for the symmetrically arranged system, and thus, the energy transfer does not occur for the one-exciton state as we can observe the curves labeled as $(1,0,0)$ and $(0,1,0)$ in Fig. 30. On the other hand, when both of QD-A and QD-B are initially excited, i.e., $(1,1,0)$, the output signal appears because three pairs of energy levels in the input and output states completely resonant. Here we emphasize that the states $|2\rangle_a|g\rangle_D$ and $|1\rangle_a|e\rangle_D$ can also couple with each other because they have same symmetries.

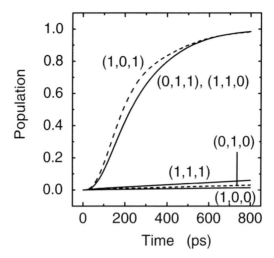

Fig. 32. Temporal evolution of the output populations for possible initial states, $(A, B, C) = (1, 0, 0)$, $(0, 1, 0)$, $(1, 1, 0)$ (without control signal), and $(1, 0, 1)$, $(0, 1, 1)$, $(1, 1, 1)$ (with control signal). The upper energy level in QD-D is adjusted equal to the energy of the other three quantum dots, and the optical near-field coupling strengths $\hbar U$ and $\hbar U'$, and the nonradiative relaxation constant Γ are set as $U = 66\,\mu\text{eV}$, $\hbar \overline{U}' = 3.3\,\mu\text{eV}$, and $(50\,\text{ps})^{-1}$, respectively

Second, we pay attention to the case that an exciton is initially prepared in QD-C. In this case, the output population only appears when either QD-A or QD-B is initially excited, which shows similar dynamics to the initial condition of $(1, 1, 0)$. The readers notice that the curves for the initial $(0, 1, 1)$ and $(1, 0, 1)$-states are slightly different. This is caused by the symmetry of the $(1, 0, 1)$-state, while the $(1, 1, 0)$-state is asymmetric. When both of QD-A and QD-B are excited, that is the three-exciton state labeled as $(1, 1, 1)$, the signal becomes small enough, which is the OFF-level in this logic gate. The above operations are summarized in Table 2. We clearly understand that the four-quantum-dot system operates as an AND- and XOR-logic gates in the cases without and with the control signal, respectively.

Such a nanophotonic device is quite interesting because two types of operations are carried out in a same quantum-dot system, which has large advantages to avoid complex nanofabrication process, and lowering the number of device elements in an integrated nanophotonic circuit.

4.4 Nanophotonic Buffer Memory

As mentioned in Sect. 4.3, we discuss realization of characteristic nanophotonic devices in the remaining sections, in which the coupling features of anti-symmetric states. Since the anti-symmetric states have no total dipole, it cannot be excited by far-field light, and also cannot radiate far-field light.

Table 2. Relationship between the input and output populations for the four-quantum-dot system

input		control	output
A	B	C	D
0	0	0	0
1	0	0	0
0	1	0	0
1	1	0	1
0	0	1	0
1	0	1	1
0	1	1	1
1	1	1	0

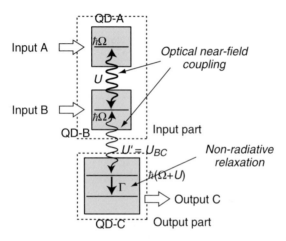

Fig. 33. Schematic illustration of maximally asymmetric arranged quantum-dot system. Two identical quantum dots, QD-A and B makes dipole inactive state for far-field light, which corresponds to the anti-symmetric

Using this dipole inactive feature, we can realize an interesting operation, in which exciton–polariton or incident photon energy is retained in the system for a long time. We refer to this type of device as photon storage or photon buffer memory.

The three-quantum-dot system is the simplest configuration to obtain a completely anti-symmetric state for the photon buffer memory, which is illustrated in Fig. 33. The upper energy level in QD-C is positively shifted by $\Delta\Omega = U$, which is the resonance condition between the input and output parts in the asymmetrically arranged quantum-dot system. For this system, the anti-symmetric state for the one-exciton state is not directly excited,

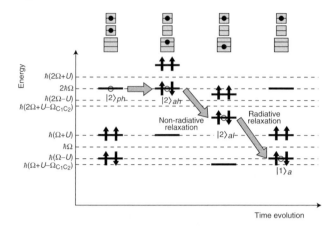

Fig. 34. Temporal sequence of energy states in a three-quantum-dot system, which starts from two-exciton state. $\hbar\Omega_{C_1,C_2}$ denotes the energy difference between upper and lower energy levels in QD-C

because both of the symmetric and anti-symmetric states are simultaneously excited. However, if the two-exciton state is initially excited in the input part, we can create completely anti-symmetric state. This is explained by using temporal sequence of the energy states as illustrated in Fig. 34. According to (88–96), the energy transfer from the two-exciton states $|2\rangle_{ph}$ to $|2\rangle_{ah}$ is active when QD-C is asymmetrically located and the upper energy level in QD-C is positively shifted by $\Delta\Omega = U$. An exciton transfers into the upper energy level in QD-C, resonantly, where the rest exciton in the coherent operation part completely stays in the anti-symmetric state. Then, the exciton on the upper energy level in QD-C drops into the lower energy level via intra-sublevel relaxation with leaving the anti-symmetric state in QD-A and QD-B. The lower energy level in QD-C is dipole active state for far-field light, therefore, the exciton annihilates due to spontaneous emission, which spends several ns, while the anti-symmetric state does not couple to far-field light because the total dipole moment in a coherent operation part is zero. Moreover, the anti-symmetric one-exciton state is off-resonant to the upper energy level in QD-C, and thus, very long lifetime is expected. In this manner, a complete anti-symmetric state for one-exciton state can be obtained in such a maximally asymmetric system.

Figure 35 represents a numerical result of the time evolution derived by using (79–96). The solid curve in Fig. 35 corresponds to the sum of exciton population for the states $|1\rangle_a$, $|2\rangle_{ah}$, and $|2\rangle_{al}$, which holds zero total dipole moment for far-field light. On the other hand, the sum of exciton population in the states, $|1\rangle_s$, $|2\rangle_{sh}$, and $|2\rangle_{sl}$ is plotted by the dashed curve in Fig. 35, which is almost zero value since the initial stage. Actually, the far-field radiation cannot be restrained completely in this system because the weak coupling from the anti-symmetric state to the symmetric state exists, where the energy

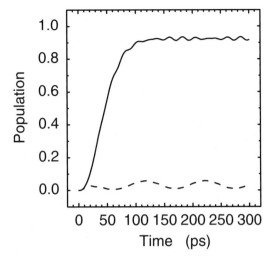

Fig. 35. Temporal evolution of exciton population for the two-exciton state. The *solid* and *dashed curves* represent for the anti-symmetric and symmetric states, respectively. The coupling strength between QD-A and B is set as $\hbar U = 89\,\mu\text{eV}$, and that between QD-B and C is $\hbar U' = \hbar U_{\text{BC}} = 14\,\mu\text{eV}$. The coupling between QD-A and C is assumed to be zero, $\hbar U_{\text{CA}} = 0$

transfer depends on the average coupling strength $\overline{U}'$ and the difference of the couplings $\Delta U'$ as shown in (79–87). The oscillating behavior is observed for both curves in Fig. 35, which is also caused by the coupling between the anti-symmetric and symmetric states. Although we have neglected the coupling to far-field light in this calculation, the slow exponential decay could be observed in the anti-symmetric state when we take into account the far-field coupling.

The above discussion is devoted to a writing process in the photon buffer memory or how to prepare an anti-symmetric state in a coupled quantum-dot system. However, in a general system, we need to consider a mechanism for reading information in addition to the writing process, which will be discussed elsewhere.

4.5 Nanophotonic Signal Splitter for Quantum Entanglement

In Sects. 4.2–4.4, we have focused on realization of conventional functional devices in nanometer space. Note that our proposed functional devices consist of two parts; coherent operation part with matter coherence and dissipative output part as illustrated in Figs. 17 or 4. The matter coherence in the coherent operation part is maintained for the period that exciton population moves to the dissipative output part. In other words, the output quantum dot acts as a selector to identify a certain quantum state by using resonant energy transfer and spatial symmetry in a quantum-dot system. Therefore, such a system is useful for detecting some information about quantum entanglement. In this

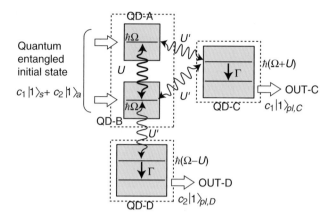

Fig. 36. Schematic illustration of signal splitter reflected quantum entangled states, which constructs from symmetrically arranged system with positive energy shift and maximally asymmetrically arranged system with negative energy shift. Two identical quantum dots, QD-A and B hold quantum mechanical motion, which is the coherent operation part, and QD-C and D correspond to the dissipative output part

section, we propose a special signal splitting device regarding to quantum entangled states. To use quantum mechanical and classical dissipative process, simultaneously, is a novel concept inherently originating from nanophotonics.

Figure 36 shows schematic illustration of the device, in which two output terminal quantum dots, QD-C and D, exist. The QD-C is located symmetrically with regard to two identical quantum dots, and the upper energy level is set with a positive shift by U. The QD-D is configured maximally asymmetrically with a negative energy shift $\Delta\Omega = -U$. From the resonance conditions, QD-C can extract exciton population from the coherent operation part via the symmetric state, while QD-D resonantly selects the anti-symmetric state, where resonant energy transfer is allowed because of the symmetry breaking (see Fig. 35). Therefore, this system can distinguish an initial quantum entangled state, $|\phi\rangle = c_1|1\rangle_s + c_2|1\rangle_a$, and information of weight coefficients c_1 and c_2 is converted to optical frequency or wavelength of far-field light, which is released after the spontaneous lifetime (several ns) of excitons.

Numerical results of the exciton population dynamics in the above system are given in Fig. 37, which is calculated by using the density matrix formalism. Figure 37a is the time evolution in the case of $|c_1|^2 = 2/3$ and $|c_2|^2 = 1/3$ as an initial condition and Fig. 37b is that of $|c_1|^2 = 1/3$ and $|c_2|^2 = 2/3$. Such initial quantum entangled states can be established by using asymmetrically located optical near-field source against QD-A and B, or by connecting to some quantum computing devices. Both cases in Fig. 37 show that the exciton population on the energy levels in QD-C and QD-D well reflects the initial weight coefficients, where the horizontal gray lines in Fig. 37 indicate the expected values. Note that the deviation from the expected values becomes large as the

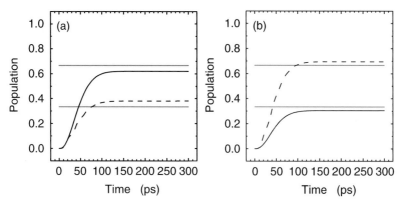

Fig. 37. Temporal evolution of exciton population on the lower energy level in QD-C (*solid curve*) and that in QD-D (*dashed curve*). (**a**) and (**b**) represent the weight coefficients $|c_1|^2 = 2/3$, $|c_2|^2 = 1/3$, and $|c_1|^2 = 1/3$, $|c_2|^2 = 2/3$, respectively. The coupling strength between QD-A and B is set as $\hbar U = 89\,\mu\text{eV}$, and that between QD-B and C (QD-A and C, QD-A and D, QD-B and D) is $\hbar U' = 14\,\mu\text{eV}$

weight coefficient of symmetric state c_1 increases. This is caused so that the off-resonant energy transfer from symmetric state to the output state slightly occurs in the asymmetrically arranged quantum-dot system. In contrast, the off-resonant energy transfer from anti-symmetric state to the output state is completely forbidden, since QD-C is located symmetrically. Therefore, a part of the exciton population that should flow into symmetrically arranged system, moves into asymmetrically arranged QD-D. This results in the difference depending on these weight coefficients.

To use quantum mechanical and classical process is one of characteristic device operations inherent to nanophotonic devices. We suppose there are further useful applications to be realized. Progressive investigation will be expected from the system architecture viewpoint.

4.6 Summary

In this section, we have discussed nanophotonic functional devices in which a coherently coupled state between optical near field and nanometric quantum dots are utilized. For a theoretical aspect, we have derived equations of motion for excitons by using the bases of appropriate coupled states, i.e., symmetric and anti-symmetric states, while we had solved them by using those of isolated quantum-dot states in Sect. 3. As a result, a principle and conditions for selective energy transfer from the coherently coupled states to the output state have been analytically shown. The selective energy transfer originates from two important factors; one is whether the coherently coupled state is symmetric or anti-symmetric, where the two states have different eigenenergies. In addition, it is a characteristic feature that optical near field can access the anti-symmetric state, which is impossible for far-field light to excite because the anti-symmetric state has dipole inactive nature. The other is that we can control exciton energy transfer by designing spatial arrangement of several quantum dots; in the case of symmetrically arranged system, anti-symmetric state is dipole inactive, while asymmetrically arranged system, the exciton population in the anti-symmetric state can be attracted intently by adjusting the system asymmetry. By using these factors and preparing a larger quantum dot for an output terminal, some functional operations can be realized. As an analytical demonstration, we have proposed AND- and XOR-logic gates, and investigated their dynamics and system tolerance due to spatial asymmetry. Consequently, the AND-logic gate can be achieved by shifting the output energy level negatively, while the XOR-gate can by shifting that positively, where the energy shifts are determined by the optical near-field coupling strength, i.e., $\pm U$. When the number of quantum dots increases, we can realize higher degree functional devices, although the selectivity or resonance conditions becomes more critical. As an example, we have proposed a controlled-type logic gate by using four quantum dots, in which the AND and XOR-logic operations are exchanged in the same device depending on the number of incident excitons.

The above logic operations have been achieved by designing the system completely symmetrically, and adjusting the output energy level to the resonance conditions. A system with spatial asymmetry is very interesting from the following viewpoints: we can create a characteristic state which decouples from far-field light. Such a state expects to have longer population lifetime than usual exciton lifetime time (spontaneous emission time), and thus, we can store photon energy or exciton–polariton energy in the system. It is useful for a buffer memory which directly stores photon energy. We have shown the operation to create a pure anti-symmetric state by using a maximally asymmetrically arranged three-quantum-dot system. For another viewpoint, we can realize an information processing device in which mixed states between symmetric and anti-symmetric states are utilized. We have proposed a device which identifies quantum entangled states by using the spatial symmetry and resonance conditions. This kind of device is different from simple quantum information processing devices. Both of quantum mechanical and classical information processing can be applied in suitable situations, which is one of inherent operations in nanophotonic devices.

In Sect. 3, although we had discussed a nanophotonic switch (equivalent to an AND-logic gate), where only energy states or resonance conditions have been considered, nanophotonic functional operations discussed in this section are realized by intentionally using the spatial degree of freedom, which is one of key features in nanophotonics. With future progress of nanofabrication techniques, where we can control materials, size, and position arbitrarily in nanometric space, nanophotonics and its device architecture have a large possibility for expanding optical and electronic device technologies. We expect that the readers notice a part of such possibility in nanophotonics from discussions in this section.

5 Conclusions

In this chapter, we have theoretically investigated operation principles of typical nanophotonic devices and their dynamics as well as formulated characteristic interaction between nanometric objects and nanometric light. The nanophotonic devices discussed here are based on several nanophotonic inherent features, such as local excited states, unidirectional energy transfer, and exciton number dependence. In Sect. 2, we have formulated energy transfer between two quantum dots by using exciton–polariton picture, and estimated strength of optical near-field coupling and energy transfer time. Moreover, we theoretically proved that the optical near-field interaction enables to excite a dipole-inactive energy level for far-field light. The energy transfer rate or optical near-field coupling strength can be expressed by the overlap integral of envelope functions for two quantum dots and spatial spreading of the optical near field which is described by a Yukawa function in the lowest perturbation. Therefore, we find that the dipole-inactive state can excite easier as inter-quantum-dot distance becomes smaller.

Using the results in Sect. 2, characteristic nanophotonic devices, which consist of several quantum dots, have been proposed in Sects. 3 and 4 and we have formulated and analyzed exciton dynamics by using quantum mechanical density matrix formalism. In Sect. 3, a nanophotonic switch has been numerically demonstrated. In the switch, we use resonant and unidirectional energy transfer, and the unidirectional energy transfer path changes due to a fermionic interaction of excitons in the lowest energy level, i.e., a state-filling. Our numerical estimation shows that the state-filling time is sub-100 ps for a CuCl quantum-cube system. Furthermore, recovery time to the initial conditions can be improved in the same order of the inverse of the optical near-field coupling by using stimulated absorption and emission process for coupling to the far-field reservoir. In other words, fast recovery time is achieved by adjusting illumination power and pulse width of control light, which is independent of spontaneous emission lifetime.

In addition to the operation principles in Sect. 3, we have proposed other kind of nanophotonic devices which adopt symmetry of exciton excited states as well as spatial arrangement of quantum dots as a novel degree of freedom. In such devices, the signal can be selectively extracted by adjusting the energy level in the output quantum dot, as the output energy level is resonant to the coherently coupled states in the input system with several quantum dots. These are useful for a multi-input computation. We have shown realization of AND and XOR-logic gates by using a symmetrically arranged three-quantum-dot system. In these device operations, it is a key feature that the resonance conditions are different depending on the exciton number. Based on the similar principle, we show possibility for a higher-degree functional device, such as a controlled-type AND- and XOR-logic gates, by using more number of quantum dots. On the other hand, resonant energy transfer via an anti-symmetric state becomes possible in an asymmetrically arranged system. Although the anti-symmetric state cannot be excited by far-field light, simultaneously, this is also the state which cannot radiate the far-field photons. We have shown that the anti-symmetric state is intentionally created by using the optical near-field coupling and its spatial asymmetry. This indicates that the photon energy can be stored in the system as the form of an exciton–polariton. In principle, a photon buffer memory can be realized by these mechanism. To extract the symmetric and anti-symmetric states selectively is equivalent to identify some quantum-entangled states, since the bases are superposition of several quantum-dot excitations. Therefore, the asymmetry of the system is useful for an interface device which manipulates quantum-entangled states. As an example, we have numerically demonstrated that quantum-entangled states with the symmetric and anti-symmetric states can be separately detected in the symmetrically and asymmetrically arranged quantum dots, which means that the quantum-entangled states are distinguished by the output photons with different energies. Although, in the field of quantum information processing, only parallel information processing is attracted, a device, which intentionally lowers the coherence or partially uses it, has large possibility to extend a viewpoint in conventional device technologies.

In the above, our discussions have been focused in the theoretical aspects. Experimental studies for nanophotonic devices have been intensively advanced. A nanophotonic switch discussed in Sect. 3 has already been verified experimentally [30]. Our theoretical study has been used for fitting into the experimental data and for evaluation of physical constant, such as the strength of optical near-field coupling. Furthermore, some interesting devices have been also proposed and investigated. For example, so-called nanofountain [42], which a nanophotonic summation device has been successfully demonstrated, and in such device, the unidirectional energy transfer is utilized. On the other hand, a device, in which coherently coupled states are used, has not been reported as far as we know. However, there are some fundamental studies related to the coherently coupled state; for example, presence of anomalous long lifetime, which originates from an anti-symmetric state, has been discussed theoretically and experimentally [43, 44]. For realization of such a device using s spatial degree of freedom in nanometric space, further progress of nanofabrication technology is required. However, nanofabrication technology has been developed recently [45, 46].

Finally, we like to spare the remaining for the future outlook. All of our investigation about nanophotonic devices are based on the quantum mechanical density matrix formalism, which is the manner to describe averaged dynamics. We anticipate that an ultimate operation of nanophotonic devices is input and output of carrier energies with single photon level. In order to realize such device, theory of optical near-field interaction must be rewritten in the form including a photon statistics rigorously, which is rather interesting problem to be solved in the near future. Furthermore, a novel possibility for nanophotonic device technology may be hidden in such consideration. Research in nanophotonics and nanophotonic device technology is still at a starting line, and we expect them further innovatively developed, which we cannot speculate.

Acknowledgements

The authors are grateful to T. Yatsui of Japan Science and Technology Agency, T.-W. Kim of Kanagawa Academy of Science and Technology, H. Hori, and I. Banno of Yamanashi University for fruitful discussions. This work was carried out at the project of ERATO, Japan Science and Technology Agency, since 1998. The authors would like to thank for the persons concerned.

References

1. M. Ohtsu, K. Kobayashi, T. Kawazoe, S. Sangu, T. Yatsui: IEEE J. Select. Top. Quant. Electron. **8**, 839 (2002)
2. A. Yariv: *Introduction to Optical Electronics.* (Holt, Rinehart and Winston, New York 1971)
3. J.W. Goodmann: *Introduction to Fourier Optics.* 2nd edn. (McGraw-Hill, Tokyo 1996)
4. *Progress in Nano Electro-Optics I–IV*, ed. by M. Ohtsu (Springer, Berlin Heidelberg New York Tokyo 2003)
5. M. Ohtsu, H. Hori: *Near-Field Nano-Optics.* (Kluwer Academic/Plenum, New York 1999)
6. T. Förster: *Modern Quantum Chemistry*, ed. by O. Sinanoğlu. (Academic Press, London 1965) pp. 93–137
7. S. Sangu, K. Kobayashi, T. Kawazoe, A. Shojiguchi, M. Ohtsu: J. Appl. Phys. **93**, 2937 (2003)
8. T. Kawazoe, K. Kobayashi, J. Lim, Y. Narita, M. Ohtsu: Phys. Rev. Lett. **88**, 067404 (2002)
9. K. Cho: *Optical Responses of Nanostructures: Microscopic Nonlocal Theory.* (Springer, Beriln Heidelberg New York Tokyo 2003)
10. C. Cohen-Tannoudji, J. Depont-Roc, G. Grynberg: *Photons and Atoms: Introduction to Quantum Electrodynamics.* (Wiley, New York 1989)
11. D.P. Craig, T. Thirunamachandran: *Molecular Quantum Electrodynamics.* (Academic Press, London 1984)
12. R.G. Woolley: *Handbook of Molecular Physics and Quantum Chemistry.* Vol. 1 (Wiley, Chichester 2003)
13. J. Knoester, S. Mukamel: Phys. Rev. A **39**, 1899 (1989)
14. J.R. Zurita-Sanchez, L. Novotny: J. Opt. Soc. Am. B **19**, 1355 (2002)
15. K. Kobayashi, M. Ohtsu: J. Microsc. **194**, 249 (1999)
16. K. Kobayashi, S. Sangu, H. Ito, M. Ohtsu: Phys. Rev. A **63**, 013806 (2001)
17. S. Sangu, K. Kobayashi, M. Ohtsu: J. Microsc. **202**, 279 (2001)

18. M. Ohtsu, K. Kobayashi: *Optical Near Fields: Electromagnetic Phenomena in Nanometric Space.* (Springer, Berlin Heidelberg New York Tokyo 2003)
19. E. Hanamura: Phys. Rev. B **37**, 1273 (1988)
20. K. Kobayashi, S. Sangu, M. Ohtsu: *Progress in Nano Electro-Optics I*, ed. by M. Ohtsu (Springer, Berlin Heidelberg New York Tokyo 2003) pp. 119–158
21. P. Fulde: *Electron Correlations in Molecules and Solids.* (Springer, Berlin Heidelberg New York 1995)
22. Y. Masumoto, M. Ikezawa, B.-R. Hyun, K. Takemoto, M. Furuya: Phys. Stat. Sol. (b) **224**, 613 (2001)
23. K. Kobayashi, T. Kawazoe, S. Sangu, M. Ohtsu: Tech. Digest of the 4th Pacific Rim Conference on Laser and Electro-Optics (2001) pp. I192–I193
24. N. Sakakura, Y. Masumoto: Phys. Rev. B **56**, 4051 (1997)
25. T. Kataoka, T. Tokizaki, A. Nakamura: Phys. Rev. B **48**, 2815 (1993)
26. H.J. Carmichael: *Statistical Methods in Quantum Optics 1* (Springer, Berlin Heidelberg New York 1999)
27. L. Mandel, E. Wolf: *Optical Coherence and Quantum Optics.* (Cambridge University, Cambridge 1995)
28. K. Akahane, N. Ohtani, Y. Okada, M. Kawabe: J. Crystal Growth **245**, 31 (2002)
29. W.I. Park, G.-C. Yi, M. Kim, S.J. Pennycook: Adv. Mater. **15**, 526 (2003)
30. T. Kawazoe, K. Kobayashi, S. Sangu, M. Ohtsu: Appl. Phys. Lett. **82**, 2957 (2003)
31. K. Lindenberg, B. West: Phys. Rev. A **30**, 568 (1984)
32. H. Hori: *Optical and Electronic Process of Nano-Matters*, ed. by M. Ohtsu. (KTK Scientific, Kluwer Academic, Tokyo Dordecht 2001) pp. 1–55
33. S. Sangu, K. Kobayashi, A. Shojiguchi, M. Ohtsu: Phys. Rev. B **69**, 115334 (2004)
34. S. De Rinaldis, I. D'Amico, F. Rossi: Appl. Phys. Lett. **81**, 4236 (2002)
35. F. Troiani, U. Hohenester, E. Molinari: Phys. Rev. B **65**, 161301 (2002)
36. E. Biolatti, R.C. Iotti, P. Zanardi, F. Rossi: Phys. Rev. Lett. **85**, 5647 (2000)
37. L. Quiroga, N.F. Johnson: Phys. Rev. Lett. **83**, 2270 (1999)
38. M.O. Scully, M.S. Zubairy: *Quantum Optics* (Cambridge Univ. Press, Cambridge 1997) pp. 222–225
39. P. Zanardi, F. Rossi: Phys. Rev. Lett. **81**, 4752 (1998)
40. M. Thorwart, P. Hänggi: Phys. Rev. A **65**, 012309 (2001)
41. B. Coffey: Phys. Rev. A **17**, 1033 (1978)
42. M. Naruse, T. Miyazaki, F. Kubota, T. Kawazoe, K. Kobayashi, S. Sangu, M. Ohtsu: Opt. Lett. **30**, 201 (2005)
43. G. Parascandolo, V. Savona: Phys. Rev. B **71**, 045335 (2005)
44. T. Kawazoe, K. Kobayashi, M. Ohtsu: IEICE Trans. Electron. **E88-C** (2005)
45. T. Yatsui, T. Kawazoe, M. Ueda, Y. Yamamoyo, M. Kourogi, M. Ohtsu: Appl. Phys. Lett. **81**, 3651 (2002)
46. W. Nomura, T. Yatsui, M. Ohtsu: Appl. Phys. Lett. **86**, 181108 (2005)

Integration and Evaluation of Nanophotonic Device Using Optical Near Field

T. Yatsui, G.-C. Yi, and M. Ohtsu

1 Introduction

Progress in DRAM technology requires improved lithography. It is estimated that the technology nodes should be down to 16 nm by the year 2019 [1]. Recent improvement of the immersion lithography using excimer laser (wavelength of 193 and 157 nm) has realized the technology node as small as 90 nm. Further decrease in the node is expected using extreme ultraviolet (EUV) light source with a wavelength of 13.5 nm. However, their resolution of the linewidth is limited by the diffraction limit of light. Furthermore, continued innovation for transistor scaling is required to manage power density and heat dissipation.

To overcome these difficulties, we have proposed nanometer-scale photonic integrated circuits (i.e., nanophotonic ICs) [2]. These devices consist of nanometer-scale dots, and an optical near field is used as the signal carrier. Since an optical near field is free from the diffraction of light due to its size-dependent localization and size-dependent resonance features, nanophotonics enables the fabrication, operation, and integration of nanometric devices.

As a representative device, a nanophotonic switch can be realized by controlling the dipole-forbidden optical energy transfer among resonant energy states in nanometer-scale quantum dots via an optical near field [3]. To realize room-temperature operation of nanophotonic switch, ZnO nanocrystallites are promising material, owing to their large exciton binding energy [4–6]. By considering the amount of the energy shift of the ground state of the exciton in the ZnO nanocrystallites due to the quantum confinement effect at room temperature, it is estimated that the size accuracy in ZnO nanocrystallites deposition must be as low as $\pm 10\%$ in order to realize efficient near-field energy transfer among the resonant energy state in nanophotonic switch composed of 5-, 7-, and 10-nm-dots [3].

In this chapter, we review the optical near-field phenomena and their applications to realize the nanophotonic device. To realize the nanometer-scale controllability in size and position, we demonstrate the feasibility of nanometer-scale chemical vapor deposition using optical near-field techniques (see Sect. 2). In which, the probe-less fabrication method for mass-production are also demonstrated. To confirm the promising optical properties of individual ZnO for realizing nanophotonic devices, we performed the near-field evaluation of the ZnO quantum structure (see Sect. 3). To drive the nanophotonic device with external conventional diffraction-limited photonic device, the far/near-field conversion device is required. Section 4 reviews nanometer-scale waveguide to be used as such a conversion device of the nanophotonic ICs.

2 Fabrication of Nanostructure Using Optical Near Field

2.1 Near-Field Optical Chemical Vapor Deposition

For realization of nanoscale photonic device required by the future system, electron beams [8] and scanning probe microscopes [9, 10] have been used to control the site on the substrate. However, these techniques have a fatal disadvantage because they cannot deal with insulators, limiting their application.

To overcome this difficulties, in this section, we demonstrated near-field optical chemical vapor deposition (NFO-CVD, Fig. 1), which enables the fabrication of nanometer-scale structures, while precisely controlling their size and position [11–16]. That is, the position can be controlled accurately by controlling the position of the fiber probe used to generate the optical near field. To guarantee that an optical near field is generated with sufficiently high efficiency, we used a sharpened UV fiber probe, which was fabricated using a pulling/etching technique [17]. In the uncoated condition, the diameter of the sharpened probe tip remained sufficiently small. This enabled high-resolution position control and in situ shear-force topographic imaging of the deposited

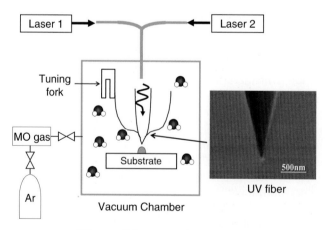

Fig. 1. Schematic of NFO-CVD

nanometer-scale structures. Since the deposition time was sufficiently short, the deposition of metal on the fiber probe and the resultant decrease in the throughput of optical near-field generation were negligible. The separation between the fiber probe and the sapphire (0001) substrate was kept within a few nanometers by shear-force feedback control. Immediately after the nanodots were deposited, their sizes and shapes were measured by in situ vacuum shear-force microscopy [11], using the same probe as used for deposition. Due to the photochemical reaction between the reactant molecules and the optical near field generated at the tip of an optical fiber probe, NFO-CVD is applicable to various materials, including metals, semiconductors, and insulators.

Conventional optical CVD method uses a light source that resonates the absorption band of megalomaniac (MO) vapor and has a photon energy that exceeds the dissociation energy [20]. Thus, it utilizes a two-step process; gas-phase photodissociation and subsequent adsorption. In this process, resonant photons excite molecules from the ground state to the excited electronic state and the excited molecules relax to the dissociation channel, and then the dissociated metallic atoms adsorb to the substrate [21]. However, we found that the dissociated MO molecules migrate on the substrate before adsorption, which limits the minimum lateral size of deposited dots (Fig. 2a). A promising method for avoiding this migration is dissociation and deposition in the adsorption-phase (Fig. 2b).

As an example of NFO-CVD, we introduce the deposition of Zn dot. Since the absorption band edge energy of the gas-phase diethylzinc (DEZn) was 4.6 eV ($\lambda = 270$ nm) [20], we used He–Cd laser light (3.81 eV, $\lambda = 325$ nm) as the light source (laser 1 in Fig. 1) for the deposition of Zn; it is nonresonant to gas-phase DEZn. However, the red-shift in the absorption spectrum in DEZn with respect to that in the gas-phase, i.e., it resonates the adsorption-phase DEZn. The red-shift may be attributed to perturbations of the free-molecule potential surface in the adsorbed-phase [20, 22]. Using a sharpened UV fiber probe, we achieved selective dissociation of adsorbed DEZn, as a results, we successfully fabricated 20-nm Zn dots with 65-nm separation on

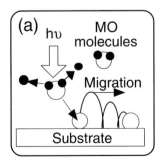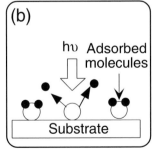

Fig. 2. Schematic diagrams of the photodissociation of the (**a**) gas-phase and (**b**) adsorption-phase MO molecules

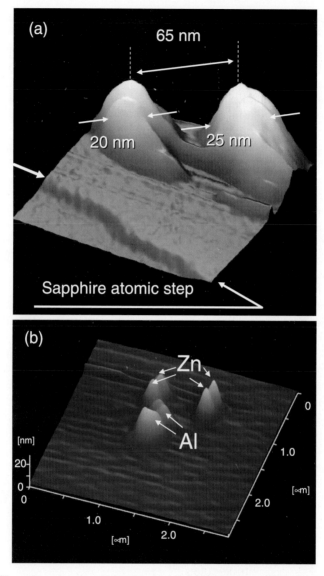

Fig. 3. Shear-force image of closely spaced (**a**) Zn dots and (**b**) Zn and Al dots

sapphire (0001) substrate (see Fig. 3a) [13, 15]. Furthermore, since the non-resonant propagating light that leaked from the probe did not dissociate the gas-phase DEZn, the atomic-level sapphire steps around the deposited dots were clearly observed after the deposition. By changing the reactant molecules during deposition, nanometric Zn and Al dots were successively deposited on the same sapphire substrate with high precision (see Fig. 3b) [16].

Since high-quality ZnO nanocrystallites can be obtained by oxidizing Zn nuclei [18, 19], NFO-CVD could be used to produce high-quality ZnO nanocrystallites; a promising material for use in nanometer-scale light-emitters and switching devices in nanophotonic IC. Furthermore, to confirm that the deposited dots were Zn, we fabricated a UV-emitting ZnO dot by oxidizing the Zn dot immediately after deposition [13].

First, the Zn dot was deposited using selective dissociation of adsorbed DEZn (Fig. 4a). Next, laser annealing was employed for this oxidization [19], i.e., the deposited Zn dot was irradiated with a pulse of an ArF excimer-laser ($\lambda = 193$ nm, pulse width: 30 ns, fluence: 120 mJ cm^{-2}) in a high-pressure oxygen environment (5 Torr). Finally, to evaluate the optical properties of the oxidized dot, the photoluminescence (PL) intensity distribution was measured using an illumination and collection mode (IC-mode) near-field optical microscope. For this measurement, an He–Cd laser ($\lambda = 325$ nm) was used as the light source and the signal collected through a fiber probe and a long wave ($\lambda > 360$ nm) pass filter was focused on a photomultiplier tube (PMT) to count photons.

Figure 4b shows the PL intensity distribution ($\lambda > 360$ nm) of an oxidized dot. The low collection efficiency due to the IC-mode configuration did not establish the spectrum. Thus, a Zn thin film was deposited using the same

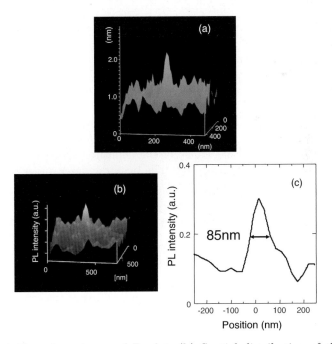

Fig. 4. (**a**) Shear-force image of Zn dot. (**b**) Spatial distribution of the optical near-field PL intensity ($\lambda > 360$ nm) of an oxidized Zn dot (**a**). The image is of a 750×750-nm area. (**c**) Cross-sectional profile through the spot of (**a**)

CVD process, except the optical near field was replaced by far-field propagating light. After it was annealed with an excimer laser, we found that the PL intensity of the spontaneous emission from the free exciton was ten times greater than that of the deep-level green emission. Thus we concluded that the PL in Fig. 4a originated from spontaneous emission from the free exciton. Figure 4c is the cross-sectional profile through the spot in Fig. 4b. Note that the full width at half maximum (FWHM) in Fig. 4c is smaller than the dot size (FWHM of 100 nm, see Fig. 4a), which was estimated using a shear-force microscope. This originates from the high spatial resolution capability of the IC-mode near-field microscope. The next stage of this study will be a more detailed evaluation of the optical properties of single ZnO nanocrystallites.

2.2 Regulating the Size and Position of Deposited Zn Nanoparticles by Optical Near-Field Desorption Using Size-Dependent Resonance

In order to realize further controllability of size, in this section, we utilize the dependence of plasmon resonance on the photon energy of optical near fields and control the growth of Zn nanoparticles during the process of Zn deposition. Using this dependence, we demonstrate the deposition of a nanometer-scale dot using NFO-CVD [14].

First, we studied nanoparticle formation on the cleaved facets of UV fibers (core diameter = 10 μm) using conventional optical CVD (see Fig. 5a).

Gas-phase DEZn at a partial pressure of 5 mTorr was used as the source gas. The total pressure, including that of the Ar buffer gas, was 3 Torr. As the light source for the photodissociation of DEZn, a 500-μW He–Cd laser (photon energy $E_p = 3.81$ eV [$\lambda = 325$ nm]) was coupled to the other end of the fiber. The irradiation time was 20 s. This irradiation covered the facet of the fiber core with a layer of Zn nanodots. Figure 6a shows a scanning electron cartographic (SEM) image of the deposited Zn nanodots, and their size distribution is shown in Fig. 6d. The peak diameter and FWHM of this curve are 110 and 50 nm, respectively.

In order to control the size distribution, we introduced 20 μW Ar⁺ ($E_p = 2.54$ eV [$\lambda = 488$ nm]) or He–Ne ($E_p = 1.96$ eV [$\lambda = 633$ nm]) lasers into the fiber, in addition to the He–Cd laser. Their photon energies are lower than the absorption band edge energy of DEZn, i.e., they are nonresonant light sources for the dissociation of DEZn. The irradiation time was 20 s. Figure 6b and c shows SEM images of the Zn nanodots deposited with irradiation at $E_p = 3.81$ and 2.54 eV and at $E_p = 3.81$ and 1.96 eV, respectively. Figure 6e and f shows the respective size distributions. The peak diameters are 30 and 18 nm, respectively, which are smaller than those of the dots in Fig. 6d, and depend on the photon energy of the additional light. Furthermore, the FWHM (10 and 12 nm, respectively) was definitely narrower than that of Fig. 6d. These results suggest that the additional light controls the size of the dots and reduces the size fluctuation, i.e., size regulation is realized.

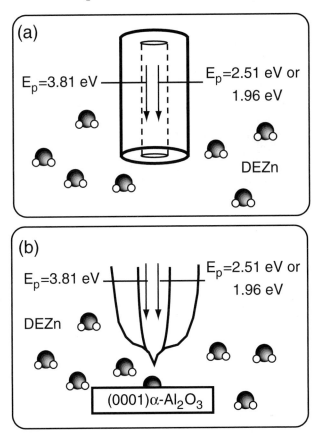

Fig. 5. Schematics of (**a**) conventional optical CVD on the cleaved facet of an optical fiber and (**b**) near-field optical CVD

We now discuss the possible mechanisms by which the additional light regulates the size of the dots. A metal nanoparticle has strong optical absorption due to plasmon resonance [24, 25], which strongly depends on particle size. This can induce the desorption of the deposited metal nanoparticles [26, 27]. As the deposition of metal nanoparticles proceeds in the presence of light, the growth of the particles is affected by a trade-off between deposition and desorption, which determines their size, and depends on the photon energy. It has been reported that surface plasmon resonance in a metal nanoparticle is red-shifted with increasing the particle size [26, 27]. However, our experimental results disagree with these reports (compare Fig. 6d–f). In order to find the origin of this disagreement, a series of calculations were performed and resonant sizes were evaluated. Mie's theory of scattering by a Zn sphere was employed, and only the first mode was considered [28]. The curves in Fig. 7a represent the calculated polarizability α with respect to three photon energies. The vertical axis is the value of α normalized to the volume, V, of a Zn sphere

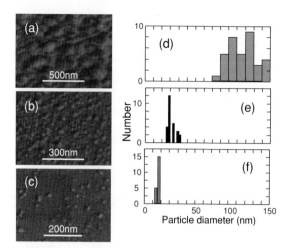

Fig. 6. SEM images of Zn nanoparticles deposited by optical CVD with (**a**) $E_p = 3.81\,eV$; (**b**) $E_p = 3.81$ and $2.54\,eV$; and (**c**) $E_p = 3.81$ and $1.96\,eV$. (**d**) The diameter distributions of the Zn particles; (**d**), (**e**), and (**f**) are for (**a**), (**b**), and (**c**), respectively

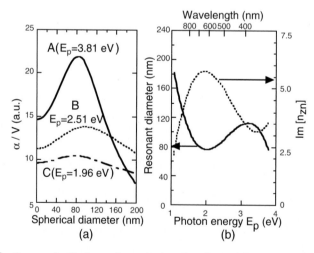

Fig. 7. (**a**) *Curves* A–C show the calculated polarizability α-normalized to the volume V for a Zn sphere surrounded by air for $E_p = 3.81$, 2.51, and $1.96\,eV$, respectively. (**b**) The resonant diameter of a Zn sphere (*solid curve*). The imaginary part of the refractive index of Zn, n_{Zn}, used for the calculation (*broken curve*) (see [29])

in the air, which depends on its diameter and is maximal at a certain diameter, i.e., at the resonant diameter. The solid curve in Fig. 7b represents the resonant diameter as a function of the photon energy, which is not a monotonous function and takes the minimum at $E_p = 2.0\,\text{eV}$ ($\lambda = 620\,\text{nm}$). Since the imaginary part of the refractive index of the Zn also takes a maximum also at $E_p = 2.0\,\text{eV}$ ($\lambda = 620\,\text{nm}$) (see the broken curve in Fig. 7b) [29], the minimum of the solid curve is due to the strong absorption in Zn.

Although Fig. 7a shows that the resonant diameter (95 nm) for $E_p = 2.54\,\text{eV}$ exceeds that (80 nm) for $E_p = 3.81\,\text{eV}$, the calculated resonant diameter for $E_p = 3.81\,\text{eV}$ is in good agreement with the experimentally confirmed particle size (see curve A in Fig. 6d). Since the He–Cd laser light ($E_p = 3.81\,\text{eV}$) is resonant for the dissociation of DEZn and is responsible for the deposition, irradiation with an He–Cd laser during deposition causes the particles to grow, and this growth halts when the particles reach the resonant diameter, because the rate of desorption increases due to resonant plasmon excitation. This is further supported by the fact that the resonant diameter (75 nm) for $E_p = 1.96\,\text{eV}$ is smaller than that for $E_p = 3.81\,\text{eV}$ (see Fig. 7a) and illumination with the additional light causes the particles to shrink (see Fig. 6d).

Another possible mechanism involves the acceleration of dissociation by the additional light. The photodissociation of DEZn produces transient monoethylzinc; then, Zn results from the dissociation of monoethylzinc. Although the absorption band of monoethylzinc was not determined, the photon-energy dependence of the size observed using the additional light might be due to the acceleration of the dissociation rate, i.e., the additional light, which is nonresonant for DEZn, resonates the monoethylzinc [30], since the first metal–alkyl bond dissociation has a larger dissociation energy than the subsequent metal–alkyl bond dissociation [31,32].

Finally, using this dependence, we used NFO-CVD (see Fig. 1) to control the position of the deposited particle. Figure 8a–c shows topographical images of Zn deposited by NFO-CVD with illumination with a 1-µW He–Cd laser ($E_p = 3.81\,\text{eV}$, laser 1 in Fig. 1) alone, or together with a 1-µW Ar$^+$ laser ($E_p = 2.54\,\text{eV}$) or a 1-µW He-Ne laser ($E_p = 1.96\,\text{eV}$) (laser 2 in Fig. 1), respectively. The irradiation times were 60 s. During deposition, the partial pressure of DEZn and the total pressure including the Ar buffer gas were maintained to 100 mTorr and 3 Torr, respectively. In Fig. 8d, curves A–C are the respective cross-sectional profiles through the Zn dots in Fig. 8a–c. The respective FWHM was 60, 30, and 15 nm; i.e., a lower photon energy gave rise to smaller particles, which is consistent with the experimental results shown in Fig. 6.

These results suggest that the additional light controls the size of the dots and reduces the size fluctuation, i.e., size regulation is realized. Furthermore, the position can be controlled accurately by controlling the position of the fiber probe used to generate the optical near field. The experimental results and the suggested mechanisms described above show the potential advantages of this technique in improving the regulation of size and position of deposited

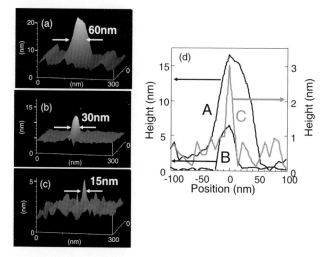

Fig. 8. Bird's-eye view of shear-force topographical images of Zn deposited by NFO-CVD with (**a**) $E_p = 3.81\,\mathrm{eV}$, (**b**) $E_p = 3.81$ and $2.54\,\mathrm{eV}$, and (**c**) $E_p = 3.81$ and $1.96\,\mathrm{eV}$, respectively. (**d**) *Curves* A–C show the respective cross-sectional profiles through the Zn dots deposited in (**a**)–(**c**)

nanodots. Furthermore, since our deposition method is based on a photodissociation reaction, it could be widely used for nanofabrication of the other material for example GaN, GaAs, and so on.

For realization of mass-production of nanometric structures, we also demonstrated possibilities of applying such a near-field desorption to other deposition technique, which does not use fiber probe. We performed metalnanoparticles deposition over the pre-formed grooves on the glass substrate (Fig. 9a) by the sputtering under the illumination (Fig. 9b). Since the optical near-field is enhanced at the edge of the groove, it can induce the desorption of the deposited metal nanoparticles when they reach at their resonant size for optical absorption (Fig. 9d and e).

By illuminating 1.96 eV (2.33 eV) light during the deposition of Au (Al) film, we successfully fabricated Au (Al) dots chain as long as 20 μm (see Fig. 10a and b).

2.3 Observation of Size-Dependent Resonance
of Near-Field Coupling Between Deposited Zn Dot and Probe
Apex During NFO-CVD

To realize sub-10-nm scale controllability in size, we report here the precise growth mechanism of Zn dots with NFO-CVD. We directly observe that the deposition rate is maximal when the dot grew to a size equivalent to the probe apex diameter. This dependence is well accounted for by the theoretically calculated dipole–dipole coupling with a Förster field. The theoretical support

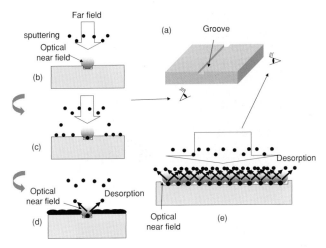

Fig. 9. Fabrication process of metal dots-chain by the spattering using near-field desorption technique

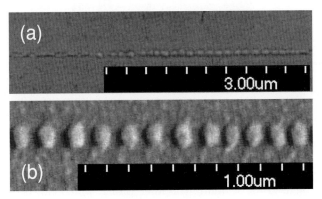

Fig. 10. (a) Au dots-chain with 1.96-eV light illumination. (b) Al dots-chain with 2.33-eV light illumination

and experimental results indicate that potential advantages of this technique for improving regulation of the size and position of deposited nanometer-scale dots.

Figure 11b shows the SEM image of the fiber probe used in this study. The apex diameter $2a_p$ is estimated as 9 nm by referring to the fitted broken circle.

Figure 12a shows a shear-force image of four Zn dots deposited with the irradiation time of 60 (dot A), 30 (dot B), 10 (dot C), and 5 s (dot D) with the laser output power P of 5 μW. We tried to deposit dots with 260 nm separations along the x-axis by servo-controlling the position of the fiber probe.

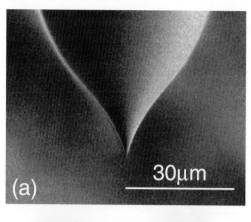

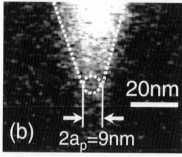

Fig. 11. A SEM image of (a) UV fiber probe and (b) its apex. $2a_p$: apex diameter

As shown in the cross-sectional profiles in Fig. 12b, Zn dots as small as 20 nm in their size S (defined by the FWHM on this profile) were fabricated. Their separation are 269 and 262 nm, by which high accuracy in position (<10 nm) was confirmed. Major origin of this residual inaccuracy is the hysteresis of the PZT actuator used for scanning the fiber probe, which can be decreased by carefully selecting the actuator.

Figure 13a shows the normalized deposition rate R of Zn dots, as a function of the dot size S. Since the measured dot size S' was convolution of probe apex diameter $2a_p$ and the real size S, S was estimated as $S = S' - 2a_p$. It should be noted that R takes the maximum at $S = 2a_p$. This result indicates that the magnitude of the near-field optical interaction between the deposited Zn dot and the probe apex is enhanced resonantly with respect to S, resulting in the resonant increase in R. In other words, the near-field optical interaction exhibits the size-dependent resonance characteristics.

To find the origin of this size-dependent resonance, we calculated the magnitude of the near-field optical interaction between closely spaced nanoparticles (Fig. 13b). The spheres 'p' and sphere 's' represent the probe apex and the Zn dot, respectively. Since the separation between two particles is much

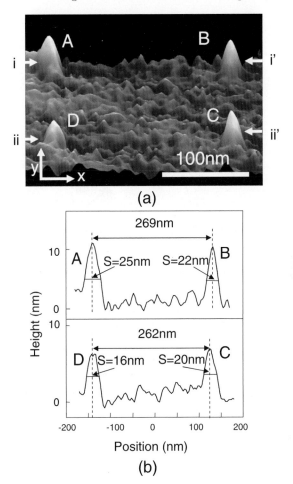

Fig. 12. (a) A shear-force image of deposited Zn dots. The laser irradiation time of dots A–D were 60, 30, 10, and 5 s, respectively. (b) *Upper* and *lower curves* show the cross-sectional profile along the line indicated by *arrows* i–i′ and ii–ii′, respectively

narrower than the wavelength, the Förster field (proportional to R^{-3}, R is the distance from the dipole) is dominant in the oscillating dipole electric field. In this quasistatic model, the intensity I_s of the light scattered from the two closely spaced spheres 'p' and 's' is given by [34]

$$I_s = I_1 + I_2 = (\alpha_p + \alpha_s)^2 |E|^2 + 4\Delta\alpha(\alpha_p + \alpha_s)|E|^2 , \qquad (1)$$

where $\alpha_i = 4\pi\varepsilon_0(\varepsilon_i - \varepsilon_0)/(\varepsilon_i + 2\varepsilon_0)a_i^3$ is a polarizability of the sphere i ($= p, s$) with diameter a_i. The first and second terms I_1 and I_2 represent the light intensity scattered from the spheres, and the light due to the dipole–dipole interaction induced by the Förster field. Thus, the light intensity under study,

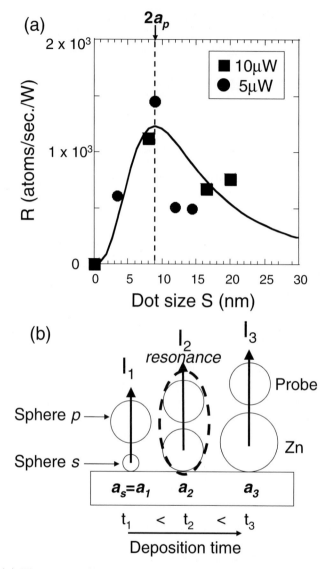

Fig. 13. (a) The time dependent deposition rate R. *Solid squares* and *circles* indicate the normalized deposition rate p with 10 and 5 μW, respectively. *Solid curve* indicates the calculated value of I_2/I_1. (b) Schematic of growth process of Zn dot

normalized to I_1, is given by

$$I_2/I_1 = \frac{G_\mathrm{p} A_\mathrm{p}^3}{(A_\mathrm{p} + 1)^3 (G_\mathrm{p} A_\mathrm{p}^3 + 1)} \,, \tag{2}$$

where $A_\mathrm{p} = a_\mathrm{p}/a_\mathrm{s}$ and $G_\mathrm{p} = (\varepsilon_\mathrm{p} - 1)(\varepsilon_\mathrm{s} + 2)/(\varepsilon_\mathrm{p} + 2)(\varepsilon_\mathrm{s} - 1)$. For the deposition by the fiber probe, dielectric constant of Zn and fiber probe are

$\varepsilon_{\rm s} = (0.6 + {\rm i}4)^2$ [35] and $\varepsilon_{\rm p} = 1.5^2$, respectively. The diameter $2a_{\rm p}$ of the sphere p was 9 nm (see Fig. 12b). Solid curve in Fig. 13a shows the calculated value of I_2/I_1 as a function of the Zn dot size S ($= 2a_{\rm s}$), which agrees very well with the experimental results. This agreement indicates that the increase in R is originated from the dipole–dipole coupling with Förster field at the dot size equivalent to the probe apex diameter.

2.4 Size-, Position-, and Separation-Controlled One-Dimensiona Alignment of Nanoparticles Using an Optical Near Field

Promising components for integrating the nanometer-sized photonic devices include chemically synthesized nanocrystals, such as metallic nanocrystals [36], semiconductor quantum dots [37], and nanorods [38], because they have uniform size, controlled shape, defined chemical composition, and tunable surface chemical functionality. However, position- and size-controlled deposition methods have not yet been developed. Since several methods have been developed to prepare nanometer-sized templates reproducibly [39], it is expected that the self-assembly of colloidal nanostructures into a lithographically patterned substrate will enable precise control at all scales [40]. Capillary forces play an important role, because colloidal nanostructures are synthesized in solution. Recently, successful integration of polymer or silica spheres [41, 42] and complex nanostructures such as nanotetrapods [42] into templates by controlling the capillary force using appropriate template structures has been reported, although their size and separation are typically uniform.

To fabricate nanophotonic devices, we propose a novel method of assembling nanoparticles by controlling the capillary force interaction and suspension flow. Further control of the positioning and separation of the nanoparticles is realized by controlling the particle–particle and particle–substrate interactions using an optical near field.

To control position and separation very accurately, preliminary experiment was performed on a patterned Si substrate, where an array of 10 μm holes in 100-nm-thick SiO$_2$ was fabricated using photolithography (Fig. 14a). Subsequently, a suspension containing latex beads with a mean diameter of 40 nm was dispersed on the substrate and the latex beads were aligned after solvent evaporation. The deposited latex beads were not subjected to any surface treatment and were dispersed in pure water at 0.001 wt%. Although the 10-μm-sized template resulted in low selectivity in the position of the latex beads (Fig. 14b and c), the beads were deposited only on the SiO$_2$ surface owing to its higher capillarity.

For higher positional selectivity, the suspension containing latex beads was dropped onto a lithographically patterned Si substrate that was spinning at 3,000 revolutions per minute (rpm). As shown in Fig. 15a, the suspension flow split into two branches at the SiO$_2$ hole. SEM images (Fig. 15b–d) show that the chain of colloidal beads was aligned at the Si/SiO$_2$ interface. Note that the number of rows of latex beads decreased (Fig. 15b and c) and only the smallest beads, which were 20 nm in diameter, reached the end of the suspension

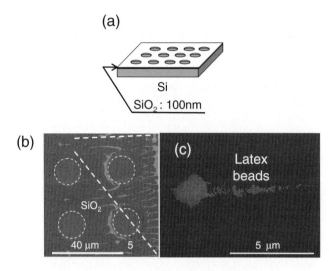

Fig. 14. (a) Schematic of lithographically patterned Si substrate. (b) and (c) SEM images of latex beads dispersed on the lithographically patterned Si substrate

flow (Fig. 15d). Assuming the same particle–suspension contact angle (denoted ψ in Fig. 15e) for various particles diameters, the flow speed of the larger latex beads had greater deceleration since the magnitude of the force pushing the particles on the SiO_2 (denoted F in Fig. 15e) owing to evaporation of the solvent is proportional to the particle diameter [42]. In other ward, the size selection was realized.

Based on the results of preliminary deposition, we tried assembling metallic nanoparticles because they are the material used to construct nanodot couplers [43]. In this trial, we investigated the assembly of colloidal gold nanoparticles with a mean diameter of 20 nm dispersed in citrate solution at 0.001%. The nanoparticles were prepared by the citric acid reduction of gold ions and terminated by a carboxyl group (approximate length is 0.2 nm) with a negative charge [44]. However, they could not be aggregated using the same deposition process as for the latex beads (Fig. 16a). To aggregate these particles, we fabricated an SiO_2 line structure with a plateau width of 50 nm on the Si substrate using photolithography (Fig. 16b). The solvent containing the colloidal gold nanoparticles was dropped onto this substrate at 3,000 rpm. Then, the colloidal gold nanoparticles aggregated along the plateau of the SiO_2 line (Fig. 16b and c). This indicates that the capillary force induced by the lithographically patterned substrate, which is caused by the higher wettability of SiO_2 than that of the Si, was larger than the repulsive force owing to the negative charge of the carboxyl group on the colloidal gold nanoparticles, and this resulted in the aggregation and alignment of the colloidal gold nanoparticles at high density.

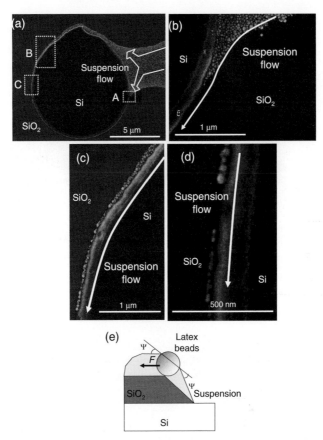

Fig. 15. (a) SEM image of latex beads dispersed on the lithographically patterned Si substrate rotated at 3,000 rpm. Higher magnification SEM images of *white squares* A (**b**), B (**c**), and C (**d**) in (**a**). (**e**) Schematic illustrating of the particle-assembly process driven by the capillary force and suspension flow

To further control size, separation, and positioning, we examined the aggregation of colloidal gold nanoparticles under illumination, because the colloidal gold nanoparticles have strong optical absorption. Strong absorption should desorb the carboxyl group from the colloidal gold nanoparticles and result in their aggregation. Such an aggregation of colloidal gold nanoparticles were confirmed by the illumination of light. Figure 17 shows the aggregated gold nanoparticles over the pyramidal Si substrate under the 690-nm-light illumination for 60 s. However, since the light was illuminated through the droplet of the colloidal gold nanoparticles, aggregated colloidal gold nanoparticles were spread outside the beam spot. In order to realize selective aggregation of the gold nanoparticles at the desired position, the suspension was illuminated from behind (Fig. 18a). Furthermore, we used an Si wedge, because this is

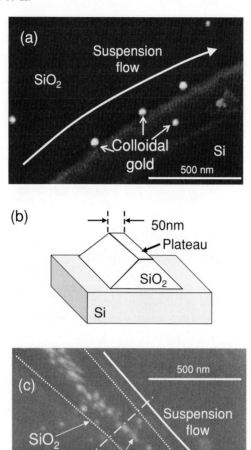

Fig. 16. (a) SEM image of colloidal gold nanoparticles dispersed on the lithograph-ically patterned Si substrate rotated at 3,000 rpm. (b) Schematic of the SiO₂ line structure fabricated on the Si substrate. (c) SEM images of colloidal gold nanopar-ticles dispersed on the SiO₂ line rotated at 3,000 rpm. *Inset*: cross-section of the substrate along the white line (*dashes* and *dots*)

a suitable structure for a far/near-field conversion device [45]. The Si wedge structure (Fig. 18b) was fabricated by the photolithography and anisotropical etching of Si. Detailed fabrication process is described in Fig. 27.

For this structure, colloidal gold nanoparticles were deposited around the edge after evaporating the suspension without illumination (Fig. 19a and b). Such aggregation is owing to its wedge structure. This is because the suspension

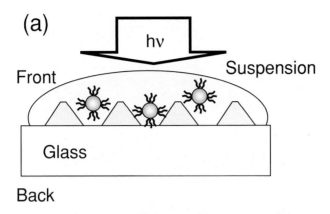

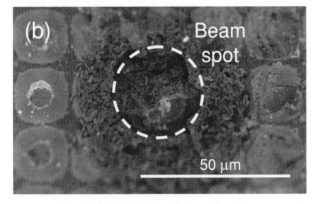

Fig. 17. (a) Frontal illumination. (b) Aggregated colloidal gold nanoparticles with frontal illumination under 690-nm light ($25\,\mathrm{mW\,mm^{-2}}$) for $60\,\mathrm{s}$

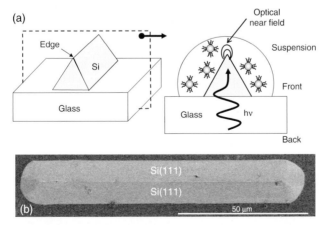

Fig. 18. (a) Schematic of the experimental setup. (b) SEM image of the fabricated Si wedge structure

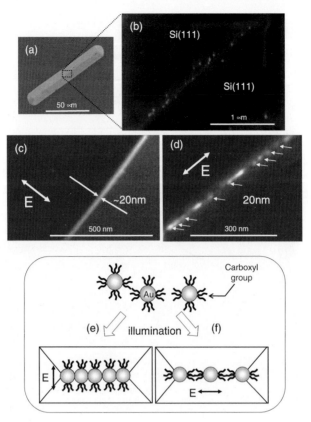

Fig. 19. (a) Overview of the Si wedge structure. (b) SEM image of colloidal gold nanoparticles deposited on the edge of the Si wedge structure without illumination. SEM images of colloidal gold nanoparticles on the Si wedge structure under illumination with polarization perpendicular (c) and parallel (d) to the edge. Schematic diagrams of the aggregation of colloidal gold nanoparticles along the edge of the Si wedge with polarization perpendicular (e) and parallel (f) to the edge

at the edge is thinner than that on the Si(111) plane owing to its low capillarity, and this causes the convective transport of particles toward the edge [46]. Further selective alignment along the edge of the Si wedge was realized using rear illumination. Figure 19c and d shows the deposited colloidal gold nanoparticles with illumination under 690-nm light ($25\,\mathrm{mW\,mm^{-2}}$) for 60 s. Since the optical near-field energy is enhanced at the edge owing to the high refractive index of Si (see Fig. 18b) [47], selective aggregation along the edge with higher density is seen in these figures. This is due to the desorption of the carboxyl group by the absorption of light by the colloidal gold nanoparticles.

Note that the colloidal gold nanoparticles were closely aggregated and aligned linearly to form a wire shape when the polarization was perpendicular to the edge axis (Fig. 19c), while they were aligned with separation of several tens of nanometers in the parallel polarization (Fig. 19d). As the optical near-field energy for parallel polarization is higher than that for perpendicular polarization [45], greater aggregation is expected for parallel polarization. Nevertheless, the parallel polarization resulted in less aggregation. The low resolution of SEM images does not determine the distribution of the carboxyl molecules. However, such a repulsive force for disaggregation is caused by the carboxyl molecules which remained on the colloidal gold nanoparticles. Thus, we believe that the difference in the degree of aggregation originated from differences in the charge distribution induced inside the gold nanoparticles. Based on the polarization dependence of the aggregation, it is reasonable to consider that the aggregation along the edge with perpendicular polarization is owing to partially adsorbed carboxyl groups (Fig. 19e), while the disaggregation with the parallel polarization resulted from the repulsive force induced by the partially attached carboxyl group on the colloidal gold nanoparticles (Fig. 19f).

3 Near-Field Evaluation of Isolated ZnO Nanorod Single-Quantum-Well Structure for Nanophotonic Device

ZnO nanocrystallites are promising material for realizing nanophotonic devices [2] owing to their large exciton binding energy [4–6], large oscillator strength [48]. Furthermore, the recent demonstration of semiconductor nanorod quantum-well structure enables us to fabricate nanometer-scale electronic and photonic devices on single nanorods because of its extremely high-quality of crystallinity [49–52]. Recently, ZnO/ZnMgO nanorod heterostructures were fabricated and the quantum confinement effect even from the SQWs was successfully observed [53]. More recently, the realization of p-type ZnO opens up significant opportunities for the opto-electro device based on ZnO [54]. Near-field spectroscopy has made a remarkable contribution to investigations of the optical properties in nanocrystallite [55], and resulted in the observation of nanometer-scale optical image, such as the local density of exciton states [56]. However, reports on semiconductor quantum structure are limited to naturally formed quantum dot (QD) [56–58]. In this section, we report low-temperature near-field spectroscopy of artificially fabricated ZnO SQWs on the end of a ZnO nanorod.

ZnO/ZnMgO SQWs were fabricated on the ends of a ZnO stem with a mean diameter of 40 nm and a length of 1 μm. They were grown vertically from the sapphire (0001) substrate using catalyst-free metalorganic vapor phase epitaxy, in which the ZnO nanorod was grown in the c orientation [52, 53]. The Mg concentration in the ZnMgO layers averaged 20 atm.%. Two samples were prepared for this study, their ZnO well layer thickness, L_w, were 2.5 and 3.75 nm, while the thicknesses of the ZnMgO bottom and top barrier

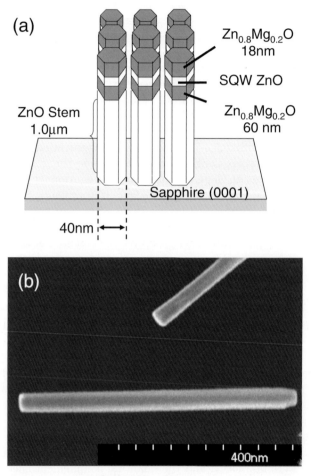

Fig. 20. (a) Schematic of ZnO/ZnMgO SQWs on the ends of ZnO nanorods. (b) SEM image of the dispersed ZnO/ZnMgO SQWs

layers in the SQWs were fixed at 60 and 18 nm, respectively. After growing the ZnO/ZnMgO nanorod SQWs, they were dispersed so that they were laid down on a flat sapphire substrate to isolate them from each other (Fig. 20).

The far-field PL spectra were obtained using an He–Cd laser ($\lambda = 325$ nm) before dispersion of the ZnO/ZnMgO nanorod SQWs. The emission signal was collected with the acromatic lens ($f = 50$ mm). To confirm that the optical qualities of individual ZnO/ZnMgO SQWs were sufficiently high, we used a collection-mode near-field optical microscope (NOM) using an He–Cd laser ($\lambda = 325$ nm) for excitation, and a UV fiber probe with an aperture diameter of 30 nm. The excitation source was focused on a nanorod sample laid on the substrate with a spot size approximately 100 μm in diameter. The PL signal was collected with the fiber probe, and detected using a cooled charge coupled device through a monochromator. The fiber probe was kept in close proximity to the sample surface ($\sim$5 nm) using the shear-force feedback technique. The polarization of the incident light was controlled with a $\lambda/2$ waveplate. In contrast to the naturally formed QD structure (a high monolayer island formed in a narrow quantum well), the barrier and cap layers laid on the substrate allowed the probe tip access to PL source, which reduced carrier diffusion in the ZnO SQWs and the subsequent linewidth broadening, thereby achieving a high spatial and spectral resolution. In addition to the PL measurements, absorption spectra were obtained using a halogen lamp, where the absorption was defined by the ratio I_{well}/I_{back} between the signal intensities transmitted through the well layer (I_{well}) and substrate (I_{back}, 50-nm apart from the well layer) (Fig. 21). The absorption signal was collected with the same fiber probe with an aperture diameter of 30 nm. Since the ZnMgO layers (bottom and top barrier layers are 60 and 18 nm, respectively) are much thicker than that of the well layer ($\sim$3 nm), any difference in the transmission signals between I_{well} and I_{back} was not detected, which resulted in no detection of the absorption peak originated from the ZnMgO layers.

As a preliminary near-field spectroscopy experiment of the ZnO SQWs, we obtained near-field PL spectra of the ZnO SQWs with $L_w = 3.75$ nm (Fig. 22a) obtained with polarization perpendicular to the c-axis ($\theta = 90$ in Fig. 20b). Two typical spectra are shown, one with a single peak at 3.375 eV (NF$_1$) and the other with several sharp peaks around 3.375, 3.444, and 3.530 eV (NF$_2$), while NF$_b$ is a background spectra (Fig. 22a). Several conclusions can

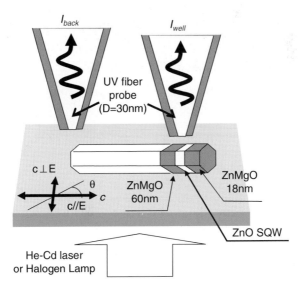

Fig. 21. Schematic of experimental setup for near-field PL spectroscopy. c: c-axis of the ZnO stem. θ: angle between the ZnO stem and the direction of incident light polarization

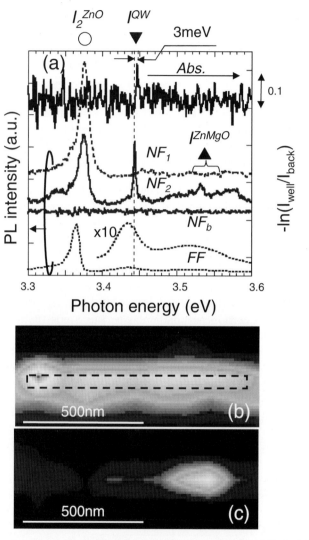

Fig. 22. Near-field PL and absorption spectroscopy of isolated ZnO SQWs (L_w = 3.75 nm) at 15 K. (**a**) NF$_1$, NF$_2$: near-field PL spectra; NF$_b$: background noise; Abs.: near-field absorption spectrum; and FF: far-field PL spectrum of vertically aligned ZnO SQWs (L_w = 3.75 nm). NOM images of isolated ZnO SQWs obtained at (**b**) 3.375 and (**c**) 3.444 eV. The rectangle shown in dashed white lines indicates the position of the ZnO stem

be drawn from these spectral profiles. First, comparison with the far-field PL spectrum (FF: dashed curve in Fig. 22a) showed that the emission peak I_2^{ZnO} at 3.375 eV was suppressed and I_{QW} (3.444 eV) and I_{ZnMgO} (3.530 eV) were enhanced in NF$_2$, indicating that peaks I_2^{ZnO} and I_{ZnMgO} originated from the ZnO stem and ZnMgO layers, respectively. Second, since the peak position of

I_{QW} was consistent with the theoretical prediction (3.430 eV) using the finite square-well potential of the quantum confinement effect in the ZnO well layer for $L_w = 3.75$ nm, we concluded that peak I_{QW} originated from the ZnO SQWs. The theoretical calculation used $0.28m_0$ and $1.8m_0$ as the effective masses of an electron and hole in ZnO, respectively, at a ratio of conduction and valance band offsets $(\Delta E_c/\Delta E_v)$ of 9, and a band gap offset (ΔE_g) of 250 meV [53]. The spatial distributions of the near-field PL intensity of peaks I_2^{ZnO} and I_{QW} (Fig. 22b and c) supported the postulate that the blue-shifted emission was confined to the end of the ZnO stem. Third, the spectral width (3 meV) of peak I_{QW} was much narrower than those of the far-field PL spectra (40 meV).

To estimate the homogeneous linewidth of isolated ZnO SQWs, we observed the power-dependence of the near-field PL spectra (Fig. 23a) by varying the excitation power densities from 0.6 to 4.8 W cm^{-2}. The shape of each spectrum was fitted using the Lorentzian function indicated by the solid curve. Figure 23b and c shows the integrated PL intensity (I_{PL}) and linewidth (Δ) of the fitted Lorentzian, which increased linearly and remained constant around 3 meV, respectively. These results indicate that emission peak I_{QW} represented the emission from a single-exciton state in ZnO SQWs and that the linewidth was governed by the homogeneous broadening. Fourth, the Stokes shift of 3 meV (Fig. 22a) was much smaller than the reported value (50 meV) in ZnO/ZnMgO superlattices [59, 60]. The small Stokes shift may result from the decreased piezoelectric polarization effect by the fully relaxed strain for the ZnO/ZnMgO nanorod quantum structures in contrast to the two-dimensional (2D) ZnO/ZnMgO heteroepitaxial multiple layers. This argument is supported by theoretical calculation of electronic structure of double barrier InAs/InP/InAs/InP/InAs nanorod heterostructures [61], concluding that any strain at heterointerfaces relaxes in nanorods within a few atomic layers in contrast to 2D pseudomorphic heteroepitaxy.

Based on these experiments, a major investigation of the optical properties of isolated ZnO SQWs was performed by analyzing the polarization-dependent PL spectrum of isolated ZnO SQWs ($L_w = 3.75$ nm). As shown in Fig. 24a, NF$_0$ is a near-field PL spectrum obtained with parallel polarization with respect to the c-axis, $\theta = 0°$, and this exhibits a new peak I_{1b}^{QW} at 3.483 eV, which is out of peak in the far-field spectrum (3.435 eV $\pm$ 20 meV). Peak I_{1a}^{QW} is the same as I_{QW} in Fig. 22a.

As the ZnO has valence band anisotropy owing to the wurtzite crystal structure, the operator corresponds to the Γ_5 (Γ_1) representation when the electric vector $\mathbf{E}$ of the incident light is perpendicular (parallel) to the crystalline c-axis, respectively. By considering the energy difference between Γ_5 and Γ_1 in the center of the zone around 40 meV for bulk material [48, 62, 63], and the direction of the incident light polarization with respect to the c-axis, emission peaks I_{1a}^{QW} and I_{1b}^{QW} in Fig. 24a are allowed for the exciton from Γ_5 and Γ_1, respectively. Note that this is the first observation of a Γ_1 exciton in a PL spectrum, while the observation of emission from Γ_1 has been realized

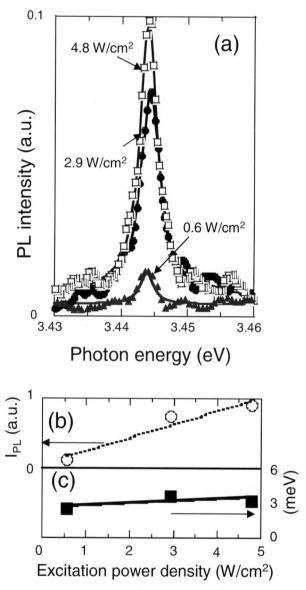

Fig. 23. Power dependent near-field PL spectroscopy of isolated ZnO SQWs ($L_w = 3.75\,\mathrm{nm}$) at $15\,\mathrm{K}$. (a) Near-field PL spectra of isolated ZnO SQWs at excitation densities ranging from 0.6 to $4.8\,\mathrm{W\,cm^{-2}}$. The integrated PL intensity, I_{PL}, (b) and homogeneous linewidths, Δ, (c) as a function of the excitation power density

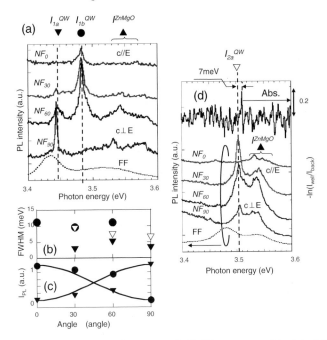

Fig. 24. Polarization-dependence of near-field PL spectra of isolated ZnO SQWs obtained at 15 K. **(a)** NF_θ, FF: near-field and far-field PL spectra of isolated ZnO SQWs ($L_w = 3.75$ nm) for $\theta = 0°$, $30°$, $60°$, and $90°$. **(b)** *Solid triangles* and *circles* are the polarization dependence of the linewidth of I_{1a}^{QW} and I_{1b}^{QW}, respectively, in (a). **(a)**. *Open triangles* are the polarization dependence of linewidth of I_{2a}^{QW} in **(d)**. **(c)** *Solid triangles* and *circles* are the integrated PL intensities, I_{PL}, of I_{1a}^{QW} and I_{1b}^{QW} , respectively, normalized to the total PL intensities ($I_{1a}^{QW} + I_{1b}^{QW}$). **(d)** NF_θ, FF: near-field and far-field PL spectra of isolated ZnO SQWs ($L_w = 2.5$ nm) and Abs.: absorption spectrum

only for bulk ZnO crystals using time-resolved reflection spectroscopy [62,63]. Since the exciton binding energy of the emission from Γ_1 (50–56 meV) [63,64] is comparable to that from Γ_5 (60 meV), this successful observation originates from the enhancement of the exciton binding energy owing to the quantum confinement effect [6]. Furthermore, in contrast to the bulk ZnO film, our sample configuration using laid ZnO nanorod SQWs has realized π polarization ($\theta = 0°$), allowing the detection of the emission from the Γ_1 exciton. The homogeneous linewidth of emission peak I_{1a}^{QW} (Γ_5) is in the range 3–5 meV, while that of I_{1b}^{QW} (Γ_1) is 9–11 meV (Fig. 24b). This difference is attributed to the degeneracy of the transition of the Γ_1 exciton with continuum and to the contribution of the residual strain field, and results in sensitive dependence of the Γ_1 exciton on the strain, as reported in the GaN [65]. The solid triangles and circles in Fig. 24c shows the respective normalized integrated PL intensity at I_{1a}^{QW} and I_{1b}^{QW}, respectively, which are in good agreement with the sine-squared and cosine-squared functions represented by the solid curves. These results indicate that emission peaks I_{1a}^{QW} and I_{1b}^{QW} originate from unidirectional transition dipoles that are orthogonal each other.

To study the linewidth broadening mechanism, Fig. 24d shows the polarization-dependent near-field PL spectra (NF$_0$–NF$_{90}$) and absorption spectrum obtained for isolated ZnO SQWs with a thinner well layer ($L_w = 2.5$ nm). In NF$_0$–NF$_{90}$, the emission peaks I_{ZnMgO} around 3.535 eV originate from the Zn-MgO layers. Emission peak I_{2a}^{QW} originates from the Γ_5 exciton in the SQWs, as was the case for I_{1a}^{QW} in Fig. 24a, since the position of peak I_{2a}^{QW} (3.503 eV) is consistent with the theoretical prediction (3.455 eV) using the finite square-well potential of the quantum confinement effect in the ZnO well layer. In comparison to ZnO SQWs with $L_w = 3.75$ nm, however, emission peak I_{2a}^{QW} had a broader linewidth (7–10 meV), which is attributed to the shorter exciton dephasing time. In the nanocrystallite where the excitons are quantized, the linewidth should be determined by the exciton dephasing time. Such dephasing arises from the collisions of the excitons at the irregular surface, so that the linewidth is d^{-2} (d is the effective size of the quantum structure) [66]. The observed well-width dependence of the spectral linewidth, $3.75^{-2}/2.5^{-2} \sim 3/7$ and the Stokes shift of 7 meV (see Fig. 24c) larger than that for $L_w = 3.75$ nm (3 meV) are supported by this dephasing mechanism quantitatively. Although emission peak I_{2a}^{QW} was suppressed for $\theta = 0°$, no peaks corresponding to the Γ_1 exciton in SQWs were detected owing to the reduction of the exciton binding energy, since the peak energy of Γ_1 for the ZnO SQWs with $L_w = 2.5$ nm is comparable with that of ZnMgO.

4 An Optical Far/Near-Field Conversion Device

For use in future photonic systems, the nanophotonic devices and ICs must be connected to conventional diffraction-limited photonic devices. This connection requires a far/near-field conversion device, such as a nanometer-scale optical waveguide. The performance parameters required of this device include:

(A) High conversion efficiency
(B) A guided beam width of less than 100 nm for efficient coupling of the converted optical near-field to sub-100 nanometer-sized dots.
(C) A propagation length that is longer than the optical wavelength to avoid direct coupling of the propagating far-field light to the nanophotonic device consisted of nanometer-scale dots. (The propagation length l_t is defined as $I(z) = I(0)\exp(-z/l_t)$, where $I(z)$ is the optical intensity and z is the longitudinal position measured from the input terminal ($z = 0$)).

4.1 A Plasmon Waveguide with Metallic-Core Waveguide

One candidate that meets these requirements is a tetrahedral tip in which the one-dimensional (1D) mode has been excited efficiently by transverse magnetic (TM)-polarized incident light [67]. Setting the incoming beam at an oblique angle to the metal slit at the edge of a tetrahedral tip converts far-field light to the 1D internal edge mode. However, the guided mode profile in a tetrahedral tip has not been observed directly. Furthermore, since it consists of a dielectric core surrounded by metal, the smallest diameter of the optical beam is estimated to be 100 nm. This is the theoretical value of the HE plasmon-mode in a cylindrical glass core (refractive index $n = 1.53$) surrounded by gold ($n = 0.17 + i5.2$) [68] at a wavelength of 830 nm (see Fig. 25c) [69].

This limitation does not meet requirement (B). In order to realize a narrower beam diameter, a more promising candidate is a cylindrical metal-core waveguide, through which the TM plasmon mode propagates [70].

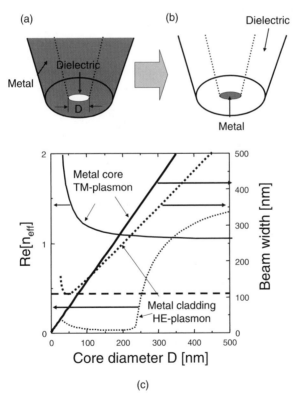

Fig. 25. Schematic of tapered (**a**) metallic-cladding and (**b**) metallic-core waveguide. (**c**) Real part of equivalent refractive index relevant guided through the waveguides and their respective beam width

This is the basis for our proposed plasmon waveguide for optical far/near-field conversion. However, the TM plasmon mode is not easily excited by far-field light, due to mode mismatching. To overcome this difficulty, we employed a metal-coated Si wedge structure. Using a near-field optical microscope, we were able to directly observe the TM plasmon mode propagating along the plateau of the Si wedge. Figure 26a shows our plasmon-waveguide scheme. The main part consists of a silicon dielectric wedge, coated with a thin metal film. Incoming far-field light, which is polarized parallel to the y-axis, is first transformed into the 2D surface plasmon mode on the F_1 side (see Fig. 26b). Next, the 2D surface plasmon mode is converted into the 1D TM plasmon mode at the edge between F_1 and the plateau. This conversion occurs because of the scattering coupling at the edge [71]. Third, the TM plasmon mode propagates along the plateau in a manner similar to surface plasmon modes using metal stripes [72] or edge modes using metal wedges [73]. This propagation occurs because the metal film deposited on the plateau is thicker than on the other faces (F_1, F_2, and F_3) due to the normal evaporation process (see Fig. 26c) [74]. Consequently, the plateau acts as a metal-core waveguide. Finally, the TM plasmon mode at the waveguide outlet is converted to the optical near field so that it couples to the nanometer-scale dots, which are located in close proximity to the outlet.

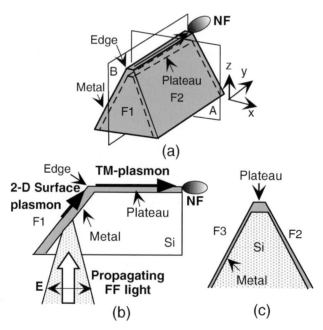

Fig. 26. (a) Bird's-eye view of a plasmon waveguide. The x and y axes are perpendicular and parallel to the plateau axis, respectively. (b) Cross-section along plane B (yz) in (b). (c) Cross-section along plane A (xz) in (b)

However, the TM plasmon mode is not easily excited by far-field light, due to mode mismatching. To overcome this difficulty, we employed a metal-coated Si wedge structure (Fig. 26a) [67]. The main part consists of an Si wedge, coated with a thin metal film. Incoming far-field light, which is polarized parallel to the y-axis, is first transformed into the 2D surface plasmon mode on the F_1 side. Next, the 2D surface plasmon mode is converted into the 1D TM plasmon mode at the edge between F_1 and the plateau. This conversion occurs because of the scattering coupling at the edge [75]. Third, the TM plasmon mode propagates along the plateau in a manner similar to edge modes using tetrahedral tip [67]. Since the metal film deposited on the plateau is thicker than that on the other faces (F_1, F_2, and F_3) due to the normal evaporation process, the plateau acts as a metal-core waveguide. Finally, the TM plasmon mode at the waveguide outlet is converted to the optical near field.

Advantages of our waveguide are:

(a) High conversion efficiency from the 2D surface plasmon mode to the 1D TM plasmon mode, due to the scattering coupling [71,75].
(b) The beam width decreases (as narrow as 1 nm) with the core diameter, since this waveguide does not have a cut-off (see Fig. 25c) [69].
(c) The propagation length of the TM-plasmon mode is sufficiently long as to meet requirement (C). For example, the propagation length is 1.25 μm (at $\lambda = 830$ nm) for a TM plasmon with a gold core (diameter $D = 40$ nm) insulated by air [69].

The plasmon waveguide was fabricated in four steps:

1. A (100)-oriented silicon wafer was bonded to the glass substrate by anodic bonding [76].
2. In order to avoid any deformation of the convex corners [77] (see Fig. 27b), the patterned rectangular mask was tilted 30° with respect to the ⟨110⟩ crystal orientation of silicon (see Fig. 27c).

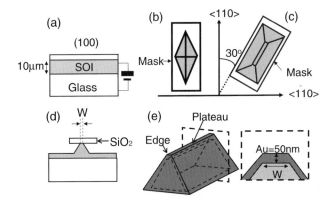

Fig. 27. Fabrication steps for a pyramidal Si probe: (**a**) anodic bonding; (**b–d**) anisotropic etching for fabrication of the Si wedge; and (**e**) metal coating

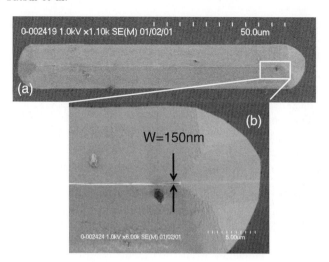

Fig. 28. SEM images of the fabricated pyramidal silicon probe: (**a**) over view; (**b**) magnified image of (**a**)

3. The Si wedge was fabricated by anisotropic etching (40 g:KOH +60 g:H$_2$O +40 g:isopropyl alcohol, 80°C) (see Fig. 27d). Maintaining the Si wedge height at less than 10 μm also kept its propagation loss sufficiently low.
4. After removing the SiO$_2$ layer, the Si wedge was coated with a 50-nm-thick gold layer (see Fig. 28a and b).

The spatial distribution of the electric-field energy throughout the plateau of metallized Si wedge was measured by scanning fiber probe with an aperture diameter D_a of 60 nm. In order to excite the plasmon mode, linearly polarized light ($\lambda = 830$ nm) was focused onto the F_1 face. Figure 29a and b shows the observed electric-field energy distributions on the wedges with plateau width $W = 1$ μm and 150 nm for TM polarization (the incident light polarization is parallel to the y). Figure 29c and d is for TE polarization (parallel to the x). Comparing Fig. 29a and c (or Fig. 29b and d) shows that the propagating mode was excited efficiently only by TM polarized incident light. The closed and open circles in Fig. 29c and d shows the cross-sectional profiles along the lines in Fig. 29a (A–A$'$ and a–a$'$) and 29b (B–B$'$ and b–b$'$). Here, transmission was defined as the ratio of the light power detected by the fiber probe to the input light power. From the dotted exponential curve in Fig. 29c fitted to the open circles, the propagation length was estimated as 1.25 μm for the 150 nm wedge. This value is comparable to the theoretical value for TM plasmon mode in a cylindrical metal-core waveguide with $D = 40$ nm and consisting of a gold core and air cladding ($\lambda = 830$ nm) [69]. From fitting the solid exponential curve in Fig. 29c to the closed circles, the propagation length for $W = 1.0$ μm was estimated as 2.0 μm, which is longer than that for $W = 150$ nm. This is because, as W increases, the effective refractive index approaches that of surface

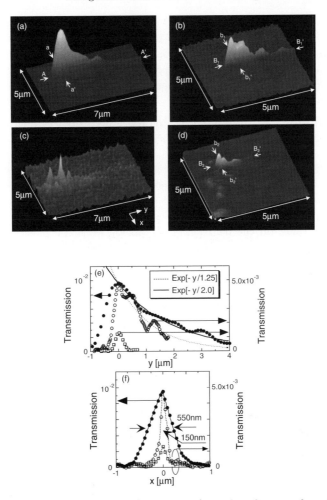

Fig. 29. Electric-field distribution ($\lambda = 830$ nm) on the silicon wedge plateau. (**a**) TM polarization: $W = 1$ μm. (**b**) TM polarization: $W = 150$ nm. (**c**) TE polarization: $W = 1$ μm. (**d**) TE polarization: $W = 150$ nm. (**e**) Cross-sectional profiles along A–A' (*closed circles*), B_1–B_1' (*open circles*), and B_2–B_2' (*open squares*) in (**a**), (**b**), and (**d**), respectively. *Solid* and *dotted curves* represent the exponential *curves* fitted to the experimental values. (**f**) Cross-sectional profiles along a–a' (*closed circles*), b_1–b_1' (*open circles*), and b_2–b_2' (*open squares*) in (**a**), (**b**), and (**d**), respectively

plasmon at the planar boundary between gold and air [69]. These experimental results confirm that the observed excitation along the plateau was the TM plasmon mode. Figure 29d shows that FWHM of the cross-sectional profiles was 150 nm for $W = 150$ nm. With minor improvements to the waveguide, this FWHM value should meet requirement (B). Furthermore, note that the transmission was 5.0×10^{-3} for $W = 150$ nm, which is ten times higher than that of a fiber probe with $D_a = 150\,\text{nm}$ [78]. This efficient excitation of the TM plasmon mode can be attributed to the scattering coupling at the edge between F_1 and the plateau in Fig. 26a [71, 75]. This transmission efficiency meets requirement (A). Finally, the propagation length estimated above is longer than the incident light wavelength. This meets requirement (C).

4.2 A Nanodot Coupler with a Surface Plasmon Polariton Condenser for Optical Far/Near-Field Conversion

To increase propagation length, a more promising candidate is a nanodot coupler consisting of an array of closely spaced metallic nanoparticles, because higher transmission efficiency is expected owing to the plasmon resonance in the closely spaced metallic nanoparticles [79]. Energy transfer along the nanodot coupler relies on the near-field coupling between the plasmon-polariton mode of the neighboring nanoparticles.

To increase the efficiency of exciting localized surface plasmon in the nanodot coupler than that excited by propagating far-field light, we equipped the nanodot coupler with a surface plasmon polariton (SPP) condenser for efficient far/near-field conversion. Figure 30 shows the proposed optical far/near-field conversion device [80].

Incoming far-field light is first transformed into the 2D SPP mode on the gold film (see Fig. 31a). Then, the SPP mode is scattered and focused by the SPP condenser, which consists of several hemispherical metallic submicron particles that are arranged in an arc and work as a "phased array" (Fig. 31b) [81]. The input terminal of the nanodot coupler is fixed at the focal point of the SPP. Finally, after the localized surface plasmon transmits through the nanodot coupler, it is converted into an optical near-field so that it couples to the nanophotonic device.

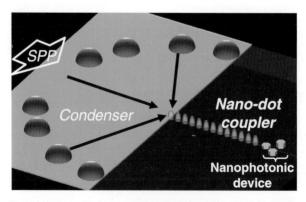

Fig. 30. Bird's eye view of a nanodot coupler with a surface plasmon-polariton (SPP) condenser

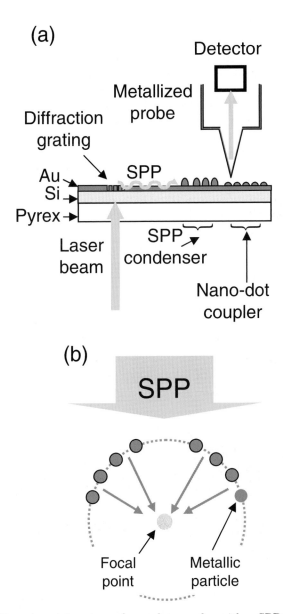

Fig. 31. (a) Experimental system of nanodot coupler with a SPP condenser. (b) Schematic illustration of the SPP condenser

The advantages of this device are that it has

(α) A high conversion efficiency, from the SPP mode to the localized surface plasmon in the nanodot coupler, owing to coupling the scattering at the inlet metallic nanoparticle [75].

(β) No cut-off diameter of the metallic nanoparticle array, i.e., the beam width decreases with the diameter because the electric field vector, which is dominant in the nanodot coupler, involves only a Förster field [79].

(γ) Longer-distance propagation than that of the metallic-core waveguide due to the reduction of metal content and plasmon resonance in the metallic nanoparticles. For example, the calculation using the finite-difference time domain (FDTD) method estimated a propagation length of $l_t = 2\,\mu m$ (at an optical wavelength $\lambda = 785\,nm$) for a plasmon-polariton mode with linearly aligned 50-nm dots with 10-nm separation [82, 83].

Advantages (α)–(γ) are compatible with meeting requirements (A)–(C), respectively.

To fabricate the nanodot coupler and SPP condenser using a focused ion beam (FIB), we used a silicon-on-insulator (SOI) wafer to avoid ion beam drift. The fabrication process was as follows:

(i) A (100)-oriented SOI wafer was bonded to a glass substrate by anodic bonding (300 V, 350°C, 10 min) (Fig. 32a) [76]. Maintaining the silicon

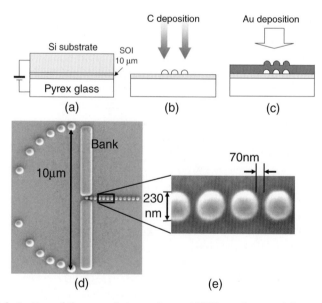

Fig. 32. Fabrication of the nanodot coupler and SPP condenser: (a) anodic bonding [step (i)], (b) carbon hemisphere deposition using FIB [step (ii)], (c) 120-nm gold film deposition using sputtering [step (iii)], (e) and (d) SEM images of the fabricated nanodot coupler and SPP condenser

wafer thickness at less than $10\,\mu m$ kept the optical propagation loss sufficiently low.

(ii) After removing the silicon substrate and SiO_2 layer from the SOI wafer by wet etching, carbon hemispheres were deposited using FIB (Fig. 32b).

(iii) To excite SPP mode and enhance the optical near-field energy, a 120-nm-thick gold film was applied using sputtering (Fig. 32c). The number of hemispheres, their positions, and their sizes were optimized using the FDTD method in order to minimize the focused spot size [83].

(iv) Finally, a diffraction grating was fabricated using an FIB milling technique $50\,\mu m$ below the condenser in order to excite the SPP by the incident optical far-field (a laser beam).

Figure 32d and e shows scanning electron microscopic (SEM) images of the SPP condenser and nanodot coupler.

Two banks were fabricated, in order to avoid illumination of the nanodot coupler by the satellite spots (originating from higher-order diffraction) (Fig.32d). The nanodot coupler consisted of a linear array of nanoparticles with diameters of 230 nm separated by 70 nm. The SPP condenser consisted of 12 scatterers 350 nm in diameter, aligned on an arc with a diameter of $10\,\mu m$.

The spatial distribution of the optical near-field energy was observed using a collection mode near-field optical microscope (see Fig.31a). A light source with a wavelength of $\lambda = 785$-nm light was used to transmit the far-field light through the 10-μm-thick Si substrate. A sharpened fiber probe with 20-nm-thick gold film was used to enhance the collection efficiency [47].

First, we checked whether the SPP condenser led to efficient scattering and resultant focusing of the SPP by exciting the SPP mode through the grating coupler. Figure 33a and b shows a shear-force image of the SPP condenser and the spatial distribution of the optical near-field energy, respectively. As shown in the cross-sectional profile (dashed curve in Fig. 33d) through the focal point of the SPP (white dotted line in Fig. 33b), FWHM of the spatial distribution of the SPP was as narrow as 400 nm. Figure 33c shows the spatial distribution of the optical near-field energy in the SPP condenser calculated using the FDTD method, where each cell was $50 \times 50 \times 25$ nm and the model consisted of $240 \times 240 \times 80$ cells. Considering the tip diameter (50 nm) of the metallized fiber probe used for collection mode, the observed distribution (Fig. 33b and dashed curve in Fig. 33d) was in good agreement with the calculated results (FWHM $= 380$ nm, solid curve in Fig. 33d). These results imply that our device works as an efficient phased array.

Second, we measured the spatial distribution of the optical near-field energy on a linear nanodot coupler, the input terminal of which was fixed at the focal point of the SPP. Figure 34a and b shows an SEM image and the spatial distribution of the optical near-field energy on the nanodot coupler, respectively. The black dots in Fig. 34c show cross-sectional profiles along the white broken line in Fig. 34b. Position $x = 0$ corresponds to the focal point

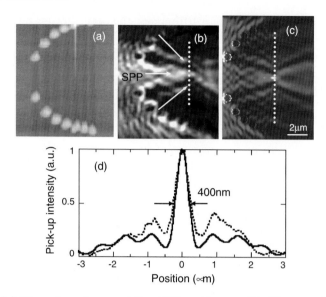

Fig. 33. (a) Shear-force image of the SPP condenser. (b) The near-field energy distribution of (a) taken at $\lambda = 785\,\text{nm}$. (c) Calculated spatial distribution of the electric-field energy using the FDTD method. The *dashed* and *solid curves* in (d) are cross-sectional profiles along the *dashed white lines* in (b) and (c), respectively

of the SPP condenser. Although not all of the energy couples to the nanodot coupler owing to mode mismatch, the optical near-field intensity has a maximum at each edge of the nanoparticles. This is due to artifact resulting from the fiber probe at constant height mode. The dips indicated by arrows A and B originate from interference of the localized surface plasmon in the nanodot coupler that arises from reflection at the output terminal. However, the exponential envelope (solid curve of Fig. 34c) fitted by neglecting these influences indicates that the propagation length l_t was $4.0\,\mu\text{m}$. l_t was five times longer than the wavelength, which meets requirement (C). The beam width was 250 nm, which is comparable to the nanoparticle size. As the size of the nanoparticles was determined by the resolution of FIB for carbon hemisphere deposition, the beam width can be decreased to sub-50 nm scale using electron beam lithography, which will meet requirement (B).

Third, we checked whether near-field coupling between the neighboring nanoparticles resulted in lower propagation loss by comparing our device with a metallic-core waveguide. For this purpose, we fabricated a gold-core waveguide the same width as the diameter of the nanoparticles in the nanodot coupler, with its input terminal also fixed at the focal point of the SPP. The open circles in Fig. 34d show the cross-sectional profile of the metallic-core waveguide and the exponential envelope (solid curve in Fig. 34d) indicates that the propagation length l_t was $1.2\,\mu\text{m}$. To evaluate the observed propagation length, we calculated the theoretical value of our metallic-core waveguide. Since the Au-core waveguide is placed on an Si substrate, we calculated

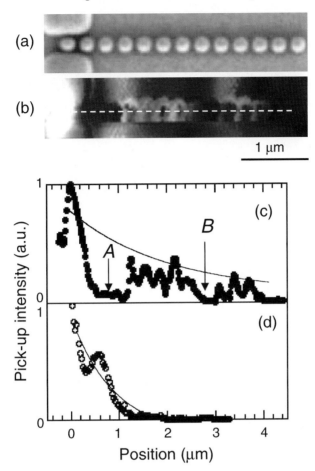

Fig. 34. SEM image (**a**) and the near-field energy distribution (**b**) of a linearly chained nanodot coupler. (**c**) *Solid circles* show the cross-sectional profiles along the *white dashed line* in (**b**). The *solid curve* shows the fitted exponential envelope. A and B indicate dips resulting from the length of the nanodot coupler. (**d**) *Open circles* show the cross-sectional profiles along metallic-core waveguide and the *solid curve* shows the fitted exponential envelope

1D TM plasmon mode [69] in the cylindrical Au-core waveguide with an diameter of $250\,\text{nm}$ ($\epsilon_{\text{Au}} = (0.174 + \text{i}4.86)^2 = -23.59 + \text{i}1.69$) [68] surrounded by the medium with an average dielectric constant of Si and air [28], $\epsilon_{\text{cl}} = ((3.705 + \text{i}0.007)^2 + 1)/2 = 7.36 + \text{i}0.03$ [68]. Based on these assumptions, the calculated propagation length of our Au-core waveguide is $1.4\,\mu\text{m}$. Since this is in good agreement with the observed value ($1.2\,\mu\text{m}$), the comparison confirmed that more efficient energy transfer was realized by the nanodot coupler.

Finally, we also observed the spatial distribution of the optical near-field energy for a zigzag-shaped nanodot coupler (see Fig. 35a and b). Corners A–D

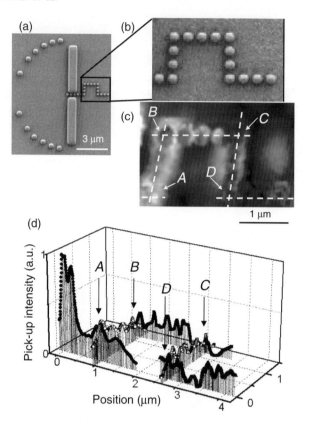

Fig. 35. SEM (**a**) and (**b**) Images and the near-field intensity distribution (**c**) of a zigzag-shaped nanodot coupler. (**d**) The cross-sectional profiles along the *dashed white lines* in (**c**). The *arrows* A–D indicate the corners

in Fig. 35c represent the profiles at locations A–D in Fig. 35d, respectively. Comparing adjacent curves, we found that the energy loss at corners A–D was negligible. This is attributed to efficient coupling of the TM and TE localized surface plasmon at the corners. As a result, the propagation length through this zigzag-shaped nanodot coupler was equivalent to that through a linear one. Although efficient coupling was predicted using the point-dipole approximation [79], this is the first experimental confirmation of it. Such high flexibility in the arrangement of nanoparticles is an outstanding advantage for optical far/near-field conversion for driving nanophotonic devices.

5 Conclusions

This chapter reviewed nanophotonics, a novel optical nanotechnology utilizing local electromagnetic interactions between a small number of nanometric elements and an optical near field. Its potential for high integration beyond the diffraction limit of light can solve the technical problems of the future optical industry.

To fabricate nanophotonic devices and ICs, we developed a new technique of chemical vapor deposition that uses an optical near field. We reported that optical near-field desorption can dramatically regulate the growth of Zn nanoparticles during optical chemical vapor deposition. The trade-off between the deposition due to 3.81-eV optical near-field light and desorption due to 2.54-eV optical near-field light allowed the fabrication of a single 15-nm Zn dot, while regulating its size and position. Furthermore, we investigated the growth mechanism of Zn dots with NFO-CVD, and found that the deposition rate was maximal when the dot grew to a size equivalent to the probe apex diameter, owing to resonant near-field coupling between the deposited Zn dot and the probe apex. The experimental results and suggested mechanisms show the potential advantages of this technique for realizing nanophotonic ICs.

As a self-assembling method for mass-production of nanophotonic ICs, we presented experimental results that demonstrate the controlled assembly of 40-nm latex particles in desired positions using capillary force interaction. Further controllability in separation and positioning was demonstrated using colloidal gold nanoparticles by introducing the Si wedge structure and controlling the direction of polarization. The experimental results and suggested mechanisms described here show the potential advantages of this technique in improving the regulation of the separation and positioning of nanoparticles, and possible application to realize a nanodot coupler for far/near-field conversion.

In order to confirm the possibility of using a nanometric ZnO dot as a light emitter in a nanophotonic IC, we report on near-field optical spectroscopy of artificially fabricated ZnO/ZnMgO nanorod SQWs as a major breakthrough for realizing nanophotonic devices using a two-level system [84, 85]. We performed both polarization-dependent absorption and emission spectroscopy of isolated ZnO nanorod SQWs, and observed valence-band anisotropy of ZnO SQWs in photoluminescence spectra directly for the first time. The success of near-field PL and absorption measurement of isolated SQWs described above is a promising step toward designing a nanophotonic switch and related devices.

To connect the nanophotonic IC with external photonic devices, we developed a nanometer-scale waveguide using a metal-coated silicon wedge structure. Propagation of the TM plasmon mode was observed using a near-field optical microscope. Illumination ($\lambda = 830\,\text{nm}$) of the metal-coated silicon wedge ($W = 150\,\text{nm}$) caused a TM plasmon mode with beam width of 150 nm and propagation length of 1.25 μm. Further improvement of the performance was realized by a nanodot coupler with an SPP condenser. The FWHM of the spatial distribution of the optical near-field energy at the focal point of the SPP was as small as 400 nm at $\lambda = 785\,\text{nm}$. Furthermore, installing a linear nanodot coupler at the focal point of the SPP realized efficient excitation of plasmon-polariton mode with a transmission length of 4.0 µm. Equivalent energy transfer was observed in zigzag-shaped nanodot couplers. These results confirm that a nanodot coupler with an SPP condenser can be used as the optical far/near-field conversion device required by future systems.

Acknowledgments

We are grateful to Wataru Nomura (University of Tokyo) and Jungshik Lim (Tokyo Institute of Technology), Drs. Tadashi Kawazoe (Japan Science and Technology Agency), Itsuki Banno (Yamanashi University), Suguru Sangu (Ricoh Company, Ltd.), and Prof. Hirokazu Hori (University of Yamanashi) for many fruitful discussions. The authors thank Messrs Jinkyoung Yoo and Sung Jin An (Pohang University of Science and Technology), and Dr. Won Il Park (Harvard University) for sample preparation of ZnO nanorod and valuable discussions.

References

1. For example, see the International Technology Roadmap for Semiconductors (http://public.itrs.net/)
2. M. Ohtsu, K. Kobayashi, T. Kawazoe, S. Sangu, T. Yatsui: IEEE J. Select. Top. Quantum Electron. **8**, 839 (2002)
3. T. Kawazoe, K. Kobayashi, S. Sangu, M. Ohtsu: Appl. Phys. Lett. **82**, 2957 (2003)
4. M.H. Huang, S. Mao, H. Feick, H. Yan, Y. Wu, H. Kind, E. Weber, R. Russo, P. Yang: Science **292**, 1897 (2001)
5. A. Ohtomo, K. Tamura, M. Kawasaki, T. Makino, Y. Segawa, Z.K. Tang, G.K.L. Wong, Y. Matsumoto, H. Koinuma: Appl. Phys. Lett. **77**, 2204 (2000)
6. H.D. Sun, T. Makino, Y. Segawa, M. Kawasaki, A. Ohtomo, K. Tamura, H. Koinuma: J. Appl. Phys. **91**, 1993 (2002)
7. D. Leonard, M. Krishnamurthy, C.M. Reaves, S.P. Denbaars, P.M. Petroff: Appl. Phys. Lett. **63**, 3203 (1993)
8. T. Ishikawa, S. Kohmoto, K. Asakawa: Appl. Phys. Lett. **73**, 1712 (1998)
9. S. Kohmoto, H. Nakamura, T. Ishikawa, K. Asakawa: Appl. Phys. Lett. **75**, 3488 (1999)
10. M. Ara, H. Graaf, H. Tada: Appl. Phys. Lett. **80**, 2565 (2002)
11. V.V. Polonski, Y. Yamamoto, M. Kourogi, H. Fukuda, M. Ohtsu: J. Microsc. **194**, 545 (1999)
12. Y. Yamamoto, M. Kourogi, M. Ohtsu, V. Polonski, G.H. Lee: Appl. Phys. Lett. **76**, 2173 (2000)
13. T. Yatsui, T. Kawazoe, M. Ueda, Y. Yamamoto, M. Kourogi, M. Ohtsu: Appl. Phys. Lett. **81**, 3651 (2002)
14. T. Yatsui, S. Takubo, J. Lim, W. Nomura, M. Kourogi, M. Ohtsu: Appl. Phys. Lett. **83**, 1716 (2003)
15. J. Lim, T. Yatsui, M. Ohtsu: IEICE Trans. Electron. E 88-C, 1832 (2005)
16. Y. Yamamoto, M. Kourogi, M. Ohtsu, G.H. Lee, T. Kawazoe: IEICE Trans. Electron. **E85-C**, 2081 (2002)

17. *Near-field nano/atom optics and technology*, ed. by M. Ohtsu. (Springer, Berlin Heidelberg New York Tokyo 1999)
18. S. Cho, J. Ma, Y. Kim, Y. Sun, G. Wong, J.B. Ketterson: Appl. Phys. Lett. **75**, 2761 (1999)
19. T. Aoki, Y. Hatanaka, D.C. Look: Appl. Phys. Lett. **76**, 3275 (2000)
20. R.R. Krchnavek, H.H. Gilgen, J.C. Chen, P.S. Shaw, T.J. Licata, R.M. Osgood, Jr.: J. Vac. Sci. Technol. B **5**, 20 (1987)
21. R.L. Jackson: J. Chem. Phys. **96**, 5938 (1992)
22. C.J. Chen, R.M. Osgood, Jr.: Chem. Phys. Lett. **98**, 363 (1983)
23. S. Yamazaki, T. Yatsui, M. Ohtsu, T.W. Kim, H. Fujioka: Appl. Phys. Lett. **85**, 3059 (2004)
24. G.T. Boyd, T. Rasing, J.R.R. Leite, Y.R. Shen: Phys. Rev. B **30**, 519 (1984)
25. A. Wokaum, J.P. Gordon, P.F. Liao: Phys. Rev. Lett. **48**, 957 (1982)
26. K.F. MacDonald, V.A. Fedotov, S. Pochon, K.J. Ross, G.C. Stevens, N.I. Zheludev, W.S. Brocklesby, V.I. Emel'yanov: Appl. Phys. Lett. **80**, 1643 (2002)
27. J. Bosbach, D. Martin, F. Stietz, T. Wenzel, F. Trager: Appl. Phys. Lett. **74**, 2605 (1999)
28. H. Kuwata, H. Tamaru, K. Miyano: Appl. Phys. Lett. **83**, 4625 (2003)
29. R.G. Yarovaya, I.N. Shklyarevsklii, A.F.A. El-Shazly: Sov. Phys.-JETP **38**, 331 (1974)
30. R.L. Jackson: Chem. Phys. Lett. **163**, 315 (1989)
31. A. Sato, Y. Tanaka, M. Tsunekawa. M. Kobayashi, H. Sato: J. Phys. Chem. **97**, 8458 (1993)
32. P.J. Young, R.K. Gosavi, J. Connor, O.P. Strausz, H.E. Gunning: J. Chem. Phys. **58**, 5280 (1973)
33. S. Cho, J. Ma, Y. Kim, Y. Sun, G. Wong, J.B. Ketterson: Appl. Phys. Lett. **75**, 2761 (1999)
34. *Optical Near Fields*, ed. by M. Ohtsu, K. Kobayashi. (Springer, Berlin Heidelberg New York 2003)
35. R.G. Yarovaya, I.N. Shklyarevsklii, A.F.A. El-Shazly: Sov. Phys. JETP **38**, 331 (1974)
36. M. Brust, C.J. Kiely: Colloids Surf. A **202**, 175 (2002)
37. A.P. Alivisatos: Science **271**, 933 (1996)
38. P. Yang: Nature **425**, 243 (2003)
39. M.D. Austin, H. Ge, W. Wu, M. Li, Z. Yu, D. Wasserman, S.A. Lyon, S.Y. Chou: Appl. Phys. Lett. **84**, 5299 (2004)
40. G.M. Whitesides, B. Grzybowski: Science **295**, 2418 (2002)
41. Y. Yin, Y. Lu, Y. Xia: J. Mater. Chem. **11**, 987 (2001)
42. Y. Cui, M.T. Björk, J.A. Liddle, C. Sönnichsen, B. Boussert, A.P. Alivisatos: Nano Lett. **4**, 1093 (2004)
43. S.A. Maier, P.G. Kik, H.A. Atwater, S. Meltzer, E. Harel, B.E. Koel, A.G. Requicha: Nature Materials **2**, 229 (2003)
44. G. Frens: Nature Phys. Sci. **241**, 20 (1973)
45. T. Yatsui, M. Kourogi, M. Ohtsu: Appl. Phys. Lett. **79**, 4583 (2001)
46. N.D. Denkov, O.D. Velev, P.A. Kralchevsky, I.B. Ivanov, H. Yoshimura, L. Nagayama: Nature **361**, 26 (1993)
47. T. Yatsui, K. Itsumi, M. Kourogi, M. Ohtsu: Appl. Phys. Lett. **80**, 2257 (2002)
48. D.C. Reynolds, D.C. Look, B. Jogai, C.W. Litton, G. Cantwell, W.C. Harsch: Phys. Rev. B **60**, 2340 (1999)

49. Y. Wu, R. Fan, P. Yang: Nano Lett. **2**, 83 (2002)
50. M.T. Björk, B.J. Ohlsson, C. Thelander, A.I. Persson, K. Deppert, L.R. Wallenberg, L. Samuelson: Appl. Phys. Lett. **81**, 4458 (2003)
51. M.S. Gudiksen, L.J. Lauhon, J. Wang, D.C. Smith, C.M. Lieber: Nature **415**, 617 (2002)
52. W.I. Park, G.-C. Yi, M.Y. Kim, S.J. Pennycook: Adv. Mater. **15**, 526 (2002)
53. W.I. Park, S.J. An, J. Long, G.-C. Yi, S. Hong, T. Joo, M.Y. Kim: J. Phys. Chem. B **108**, 15457 (2004)
54. A. Tsukazaki, A. Ohtomo, T. Onuma, M. Ohtani, T. Makino, M. Sumiya, K. Ohtani, S.F. Chichibu, S. Fuke, Y. Segawa, H. Ohno, H. Koinuma, M. Kawasaki: Nature Materials **4**, 42 (2005)
55. K. Matsuda, T. Saiki, S. Nomura, M. Mihara, Y. Aoyagi: Appl. Phys. Lett. **81**, 2291 (2002)
56. K. Matsuda, T. Saiki, S. Nomura, M. Mihara, Y. Aoyagi, S. Nair, T. Takagahara: Phys. Rev. Lett. **91**, 177401 (2003)
57. J.R. Guest, T.H. Stievater, G. Chen, E.A. Tabak, B.G. Orr, D.G. Steel, D. Gammon, D.S. Katzer: Science **293**, 2224 (2001)
58. T. Guenther, C. Lienau, T. Elsaesser, M. Clanemann, V.M. Axt, T. Kuhn, S. Eshlaghi, D. Wieck: Phys. Rev. Lett. **89**, 057401 (2002)
59. A. Ohtomo, M. Kawasaki, I. Ohkubo, H. Koinuma, T. Yasuda, Y. Segawa: Appl. Phys. Lett. **75**, 980 (1999)
60. T. Makino, A. Ohtomo, C.H. Chia, Y. Segawa, H. Koinuma, K. Kawasaki: Physica E **21**, 671 (2004)
61. M. Zerovos, L.-F. Feiner: J. Appl. Phys. **95**, 281 (2004)
62. M. Zamfirescu, A. Kavokin, B. Gil, G. Maplpuech, M. Kaliteevski: Phys. Rev. B **65**, 161205 (2001)
63. S.F. Chichibu, T. Sota, G. Cantwell, D.B. Eason, C.W. Litton: J. Appl. Phys. **93**, 756 (2003)
64. D.C. Reynolds, C.W. Litton, D.C. Look, J.E. Hoelscher, C. Claflin, T.C. Collins, J. Nause, B. Nemeth: J. Appl. Phys. **95**, 4802 (2004)
65. M. Tchounkeu, O. Briot, B. Gil, J.P. Alexis, R.L. Aulombard: J. Appl. Phys. **80**, 5352 (1996)
66. T. Wamura, Y. Masumoto, T. Kawamura: Appl. Phys. Lett. **59**, 1758 (1991)
67. U.C. Fischer, J. Koglin, H. Fuchs: J. Microsc. **176**, 281 (1994)
68. *Handbook of Optical Constants of Solids*, ed. by E.D. Palik. (Academic, New York 1985)
69. L. Novotny C. Hafner: Phys. Rev. E **50**, 4094 (1994)
70. J. Takahara, S. Yamagishi, H. Taki, A. Morimoto, T. Kobayashi: Opt. Lett. **22**, 475 (1997)
71. *Light Transmission Optics*, ed. by D. Marcuse. Chap. IV (Van Nostrand Reinhold, New York 1972)
72. J.C. Weeber, J.R. Krenn, A. Dereux, B. Lamprecht, Y. Lacroute, J.P. Goudonnet: Phys. Rev. B **64**, 045411 (2001)
73. A. Eguiluz, A.A. Maradudin: Phys. Rev. B **14**, 5526 (1976)
74. K. Tachibana, T. Someya, S. Ishida, Y. Arakawa: Appl. Phys. Lett. **76**, 3212 (2000)
75. T. Yatsui, M. Kourogi, M. Ohtsu: Appl. Phys. Lett. **71**, 1756 (1997)
76. T.R. Anthony: J. Appl. Phys. **58**, 1240 (1998)
77. B. Puers, W. Sansen: Sens. Actuat. **A21-A23**, 1036 (1990)

78. T. Yatsui, M. Kourogi, M. Ohtsu: Appl. Phys. Lett. **73**, 2090 (1998)
79. M.L. Brongersma, J.W. Hartmanm, H.A. Atwater: Phys. Rev. B **62**, R16356 (2000)
80. W. Nomura, M. Ohtsu, T. Yatsui: Appl. Phys. Lett. **86**, 181108 (2005)
81. I.I. Smolyaninov, D.L. Mazzoni, J. Mait, C.C. Davis: Phys. Rev. B **56**, 1601 (1997)
82. M. Quinten, A. Leitner, J.R. Krenn, F.R. Aussenegg: Opt. Lett. **23**, 1331 (1998)
83. The computer simulations in this paper are performed by a FDTD-based program, Poynting for Optics, a product of Fujitsu, Japan
84. A. Zenner, E. Beham, S. Stufler, F. Findeis, M. Bichler, G. Abstreiter: Nature **418**, 612 (2002)
85. Z. Yuan, B.E. Kardynal, R.M. Stevenson, A.J. Shields, C.J. Lobo, K. Cooper, N.S. Beattie, D.A. Ritchie, M. Pepper: Science **295**, 102 (2002)

Unique Properties of Optical Near Field and their Applications to Nanophotonics

T. Kawazoe, K. Kobayashi, S. Sangu, M. Ohtsu, and A. Neogi

1 Introduction

The optical science and technology of the 21st century requires an optical nanotechnology that goes beyond the diffraction limit. To meet this requirement, Ohtsu [1] has proposed a novel technology called nanophotonics. Nanophotonics is defined as a technology that utilizes local electromagnetic interactions among small nanometric elements via an optical near field. Since an optical near field is free from light diffraction due to its size-dependent localization and resonance features, nanophotonics enables the fabrication, operation, and integration of nanometric devices. The primary advantage of nanophotonics is its capacity to realize novel functions based on local electromagnetic interactions. It should be noted that some of the conventional concepts of wave optics, such as interference, do not apply in nanophotonics. Instead, key concepts include surface elementary excitation and nanofabrication technology. This chapter introduces the unique properties of optical near fields, which were obtained experimentally, and their applications to nanophotonic devices and nanofabrications. Section 1 outlines the chapter contents, introducing each theme based on features of the optical near field. Sections 2–7 review experimental results and discussion.

1.1 Excitation Energy Transfer via Optical Near Field

Observations of a single quantum dot (QD) have recently become possible using near-field optical spectroscopy [2,3] and microluminescence spectroscopy [4, 5]. Spectroscopy of individual QDs is one of the most important topics of nanostructure physics, and several extraordinary phenomena have been reported, such as intermittent luminescence [6] and long coherent time [7]. The coupled QDs system exhibits more unique properties (e.g., the Kondo effect [8,9], Coulomb blockade [10,11], and breaking the Kohn theorem [13]) in contrast with the single QD system. The QDs system reveals a variety of interactions such as carrier tunneling [8–10, 12], Coulomb coupling [11], spin interaction [13], and so on. Investigation of interactions among QDs is important not only for deep understanding the various physical phenomena but also for developing of novel functional devices [8–13]. The optical near-field interaction [1] is of particular interest, as it can govern the coupling strength among QDs.

Mukai et al. [14] reported ultrafast "optically forbidden" energy transfer from an outer ring of loosely packed bacteriochlorophyll molecules, called B800, to an inner ring of closely packed bacteriochlorophyll molecules, called B850, in the light-harvesting antenna complex of photosynthetic purple bacteria. They theoretically showed that this transfer is possible when the point transition dipole approximation is violated due to the size effect of B800 and B850. From our viewpoint, this energy transfer is due to the optical near-field interaction between B800 and B850. Similarly, the energy can be transferred from one dot to another by the optical near-field interaction for the QDs system, even if it is a dipole-forbidden transfer.

The energy transfer from smaller to larger QDs have been studied by the spectrally, spatially, and time resolved experiment. Kagan et al. [15] observed the energy transfer among CdSe QDs coupled by dipole–dipole interdot interaction. We proposed the model for the unidirectional resonant energy transfer between QDs via optical near-field interaction and observed the optically forbidden energy transfer among randomly dispersed CuCl QDs using the optical near-field spectrometer [16]. The theoretical analysis and time evolution of the energy transfer via optical near-field interaction have also been discussed [17, 18]. Crooker et al. [19] studied the dynamics of exciton energy transfer in close-packed assemblies of monodisperse and mixed-size CdSe nanocrystal QDs, and reported the energy-dependent transfer rate of excitons from smaller to larger dots. Section 2 reviews our work that is the direct observation of energy transfer from the exciton state in a CuCl QD to the optically forbidden exciton state in another CuCl QD by optical near-field spectroscopy.

1.2 Nanophotonic Switch Using Energy Transfer among QDs

Optical fiber transmission systems require increased integration of photonic devices for higher data transmission rates. It is estimated that the size of photonic matrix switching devices should be reduced to a subwavelength scale, as in the near future it will be necessary to integrate more than $10,000 \times 10,000$ input and output channels on a substrate [1]. Since conventional photonic devices, e.g., diode lasers and optical waveguides, have to confine light waves within their cavities and core layers, respectively, their minimum sizes are limited by the diffraction of light [20]. Therefore, they cannot meet the size requirement, which is beyond this diffraction limit. An optical near field is free from the diffraction of light and enables the operation and integration of nanometric optical devices. Namely, by using a localized optical near field as the carrier, which is transmitted from one nanometric element to another, the above requirements can be met. Based on this idea, we have proposed nanometer-sized photonic devices, that is *Nanophotonic Device* [1]. Several kinds of nanophotonic devices, which we have proposed, operate using this phenomenon.

A nanometric all-optical switch (i.e., a nanophotonic switch) is one of the most important devices for realization of a nanophotonic integrated circuits. We have succeeded in the operation of the nanophotonic switch using a coupled QD system. In Sect. 3, we report the demonstration of repetitive operation of a nanophotonic switch, using three CuCl QDs [21, 22]. The optically forbidden energy transfer between neighboring CuCl QDs via optical near-field interaction, which is reviewed in Sect. 2, is a key phenomenon for this operation.

1.3 Optical Nanofountain: Biomimetic Device

Although the internal efficiency of nanophotonic devices is very high [21, 22], for efficient operation of the system an inter-connection device needs to be developed to collect the incident propagating light and drive the nanophotonic device [1, 23]. Conventional far-field optical devices, such as convex lenses and concave mirrors, cannot be used for this inter-connection because of their diffraction-limited operation. In Sect. 4, we explain and demonstrate an optical device that we call the *optical nanofountain*, which concentrates optical energy in a nanometric region by using optical near-field energy transfer among QDs. This nanometric optical device enables not only highly efficient inter-connection to nanophotonic devices but also other nanometric optical operations and measurements, e.g., nanometric optical tweezers, highly sensitive nanometric resolution microscopes, and so on.

As we and other research groups had mentioned [16,19], the principle of the energy transfer among QDs is equivalent to that of the light-harvesting photosynthetic system, which skillfully concentrates and harvests optical energy in nanometric photosynthetic systems. Figure 1 shows the schematic explanation of the photosynthetic purple bacteria *Rhodopseudomonas acidophila* [24, 25] which has two light-harvesting antennae: LH1 and LH2. LH1 contains a 32-bacteriochlorophyll (BChl) ring, and LH2 contains a B800 ring with 9 BChls and a B850 ring with 18 BChls. They harvest photons and efficiently transfer the excitation energy from B800 to LH1, where the excitonic energy of B800 is higher than that of LH1. This unidirectional energy transfer is due to the nanometric dipole–dipole interaction, i.e., an optical near-field interaction [26], among BChl rings with low energy dissipation [14]. The optical energy concentrator *optical nanofountain*, which we propose, operates in the same manner as the light harvesting system in the photosynthetic bacteria.

1.4 Laterally Coupled GaN/AlN Quantum Dots

The successful development of short wavelength lightemitting diodes and the most recent realization of nitride-based QD lasers have stimulated great

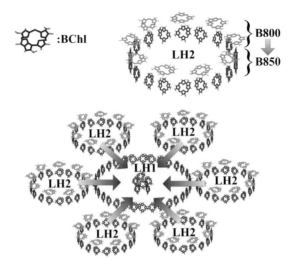

Fig. 1. Schematic explanation of the photosynthetic purple bacteria *Rhodopseudomonas acidophila*

interest in the application of quantum confined structures for blue and ultraviolet optoelectronic devices [27–29]. In particular, III-nitride-based self-assembled QDs are very promising for a wide range of commercial applications including the nanophotonic device [1,30–33]. The study of self-assembled GaN QDs presents a challenge, as the placement of individual dots is difficult to control during the epitaxial growth process, and the dot density can be quite high [34,35]. Thus, traditional experimental techniques often only allow simultaneous observation of large ensembles of QDs where inhomogeneous broadening washes out many of the interesting features. We have investigated the optical properties of GaN QD and have observed that the built-in strain fields significantly influence the radiative recombination lifetime [36–38]. The role of size distribution of the QDs on the radiative emission process is not yet clear [38].

Reports on the near-field optical properties of GaN QD studied using illumination mode are limited by the low spatial resolution due to carrier diffusion accentuated by a large dot size inhomogeneity [39,40]. The contribution from individual dots or coupled QD clusters exhibiting narrow near-field photoluminescence (PL) line shape ($\sim$ few meV) from high-spatial resolution is yet to be reported. The PL line shape of individual dots in the GaN system is expected to be significantly broader than GaAs- or InP-based QDs due to broadening induced by a significantly larger LO phonon scattering rate. In Sect. 5, we present the near-field optical emission characteristics from a cluster of a few GaN QDs with very high spatial resolution and also discuss the lateral and vertical electronic coupling of dots caused by interdot scattering of carriers.

1.5 Nonadiabatic Nanofabrication

We have studied the application of optical near field to nanofabrication, by applying the novel properties of optical near field to photochemical reactions, and have demonstrated the feasibility of the chemical vapor deposition (CVD) of Zn dots using optical near-field techniques [41–43]. So far, we have used the high spatial resolution capability of optical near field to deposit 20-nm-wide Zn wires [41] and 25-nm Zn dots [43].

Conventional optical CVD utilizes a two-step process; photodissociation and adsorption. For photodissociation, far-field light must resonate the reacting molecular gasses in order to excite molecules from the ground state to an excited electronic state [44,45]. The Franck–Condon principle holds that this resonance is essential for excitation [44]. The excited molecules then relax to the dissociation channel, and the dissociated Zn atoms adsorb to the substrate surface. For near-field optical CVD (NFO-CVD), photodissociation can take place even under nonresonant conditions. For example, we succeeded in the photodissociation of metal organic molecules and the deposition of Zn dots using a nonresonant optical near field with a photon energy lower than the energy gap of the electronic state of the molecule [46]. This photochemical reaction comes from nonadiabatic photochemical process and is one of the unique phenomena of optical near field. In addition to optical CVD, these phenomena are applicable to many other photochemical nanotechnologies. Therefore, it is very important to clarify their physical origin. In the first half of Sect. 6, we demonstrate the nonadiabatic NFO-CVD of Zn nanodots and show the incident optical-power and photon-energy dependencies of the deposition rate of Zn. Finally, we explain the experimental results based on the features of ONF and the exciton–phonon polariton (EPP) model.

Novel methods of nanofabrication are required for the mass-production of photonic and electronic devices. Fabrication using nanoimprint lithography, near-field phase mask lithography, and evanescent near-field optical lithography (ENFOL) [47] is less expensive and has a higher throughput than that using electron or ion beam, X-ray, or deep UV lithography. ENFOL is especially useful because conventional photolithographic components and systems can be used.

The spatial locality of the optical near field can be used for novel fabrication. For example, we have found that with NFO-CVD, metal organic molecules are photodissociated by the optical near field even with a lower photon energy than the dissociation energy of the molecule [46]. This unique photodissociation is explained by a nonadiabatic photochemical process based on the exciton–phonon polariton model [48,49]. According to this model, the nonadiabatic photochemical process is a universal phenomenon and is applicable to many other photochemical processes, including the exposure of photoresist.

Based on this consideration, we proposed a novel photolithography method using the nonadiabatic photochemical process, i.e., nonadiabatic near-field photolithography. Using this method, the UV-photoresist, which is suitable for nanolithography, can be exposed by cheaper visible light sources and equipments without expensive UV light sources. In the latter half of Sect. 6, we review the nonadiabatic near-field photolithghraphy.

In the nanometric photolithography, problems, originated from the wave properties of light, are not only the diffraction limit but also coherency and polarization dependence. For the photolithography of high density nanometric arrays, the optical coherent length is longer than the separation between the adjacent corrugations, even when the Hg lamp was used, and the absorption of photoresist is not enough to suppress the interference fringes of the scattered light due to their narrow separation. The transmission intensity of light passing through the photomask strongly depends on its polarization. Thus, the design of the photomask structures has to include such effects. In the nonadiabatic near-field photolithography, since the optical near-filed has no wave properties, these problem are easily solved.

1.6 Control of an Optical Near Field Using a Fiber Probe

In Sect. 7, we review the demonstrations to control the optical near filed using fiber probe by itself, which include the optical near-field features and are applicable to the nanophotonic devices. Near-field optics offers a unique technique of high spatial resolution beyond the diffraction limit of light [1]. Various interesting studies, such as spectroscopy of QDs [50] and polymers, NFO CVD [41–43, 46, 48], and so on, have been carried out by using this technique. Second-harmonic (SH) NFO imaging of a metal surface has been also reported [51]. Second harmonic generation (SHG) is a useful phenomenon not only for far-field spectroscopy but also NF spectroscopy to investigate material properties and to exploit nonlinear optical devices. SH signal is generated by the nonlinear polarization under the asymmetric condition of the material. A fiber probe used for NF optics can satisfy this condition because it has large asymmetric regions between the sharpened fiber core, coated metal, and air. In addition, the coated metal surface can enhance the SHG intensity due to the surface plasmon effect. Smolyaninov et al. [51] pointed out that the SHG efficiency in the apertureless metallic tip may not be so low in contrast with that of a bulky nonlinear crystal. Also in the metal coated apertured fiber probe, it may not be so low because of the complicated profiles of its boundary between glass and metal. However, investigation of SHG in the fiber probe has not yet been reported in spite of its potential application to frequency conversion. Thus, we believe that it is important to investigate the SHG phenomenon in the fiber probe. In the first half of Sect. 7, we demonstrate SHG in the fiber probe and review its several properties.

Optical fiber and guided-wave techniques have brought high efficiency, function, and downsizing to optical devices. The demand for downsizing is particularly strong, and several approaches, such as photonic crystals [52] and nanophotonics [1] have already been reported. Control of polarization in such devices is an important function with regard to the optical isolator, circulator, and so on. To apply polarization of light to nanophotonic devices, it is necessary to control it. However, current polarization control devices are too large to integrate in downsizing, because they require several wavelength sizes to use optical anisotropy. Optical anisotropy can be established by using structure instead of anisotropy of the optical crystal. There have been several recent studies about the polarization of optical near fields [53,54]; these used the structure of optical near-field probes and asymmetric nanodots to calculate and demonstrate selection of the polarization of an optical near field. In this study, we proposed another approach to control the polarization of light in a nanometric region. The magneto-optical effect of Fe is more than 10^4 times greater than that of glass. For a thin Fe film 1 μm thick, the degree of polarization and the Faraday rotation angle of transmitted light are more than 0.9 and 30°, respectively, at saturation magnetic flux density [55,56]. However, the depth of Fe penetration is less than 100 nm for light with a wavelength of 600 nm, so the large magneto-optical effect of Fe is unsuitable for conventional photonic devices. On the other hand, since nanophotonic devices are smaller than the penetration depth of Fe [1,57], the optical loss is not a serious problem. It is probable that the optical properties of Fe are necessary, rather than unsuitable, for nanophotonics, because the large magneto-optical effect provides one solution to the difficulty of controlling light polarization in the nanometric region. In order to develop a nanophotonic polarization controller that uses the magneto-optical effect of Fe, we prepared an optical near-field fiber probe coated with Fe [58]. In the tip of the probe, the optical field passes through a waveguide, which is narrower than the light wavelength and is covered with Fe, as shown in Fig. 2. The polarization of light passing

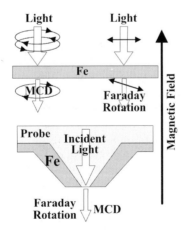

Fig. 2. Schematic explanation of the strong magneto-optical effect of Fe and an optical near-field probe coated with Fe

the tip changes owing to the strong magneto-optical effect of the Fe coating, and can be controlled using the intensity of the external magnetic field. In the latter half of Sect. 7, we show a novel approach to control the polarization of light, and demonstrate the great magnetic circular dichroism and Faraday rotation in a nanometric region using optical near-field probes coated with Fe and Fe/Au/Fe/Au.

2 Excitation Energy Transfer via Optical Near Field

2.1 Optically Forbidden Energy Transfer between CuCl QDs

Cubic CuCl QDs embedded in an NaCl matrix have the potential to be an optical near-field coupling system that exhibits this optically forbidden energy transfer. This is because, for this system, other possibility of energy transfer, such as carrier tunneling, Coulomb coupling, and so on, can be neglected, because carrier wave function is localized in QDs due to a deep potential depth of more than 4 eV and the Coulomb interaction is weak due to small exciton Bohr radius of 0.68 nm in CuCl [16]. Conventional optical energy transfer is also negligible, since the energy levels of nearly perfect cubic CuCl QDs are optically forbidden, and are confined to the energy levels of exciton with an even principal quantum number [59]. However, Sakakura et al. [60] observed the optically forbidden transition in a hole-burning experiment using cubic CuCl QDs. Although they attributed the transition to an imperfect cubic shape, the experimental and simulation results did not show such imperfection. We believe that the transition was due to the energy transfer between the CuCl QDs, via the optical near-field interaction similar to the optically forbidden energy transfer between B800 and B850 in the above-mentioned photosynthetic system. Thus far, this type of energy transfer has not been directly observed.

It is well known that translational motion of the exciton center of mass is quantized due to the small exciton Bohr radius for CuCl QDs, and that CuCl QDs become cubic in an NaCl matrix [60–62]. The potential barrier of CuCl QDs in an NaCl crystal can be regarded as infinitely high, and the energy eigenvalues for the quantized Z_3 exciton energy level (n_x, n_y, n_z) in a CuCl QD with the side length of L are given by

$$E_{n_x, n_y, n_z} = E_{\mathrm{B}} + \frac{\hbar^2 \pi^2}{2M(L - a_{\mathrm{B}})^2}(n_x^2 + n_y^2 + n_z^2),\tag{1}$$

where E_{B} is the bulk Z_3 exciton energy, M the translational mass of exciton, a_{B} its Bohr radius n_x, n_y, and n_z quantum numbers $(n_x, n_y, n_z = 1, 2, 3, \ldots)$, and $d = (L - a_{\mathrm{B}})$ corresponds to an effective side length found through consideration of the dead layer correction [60]. The exciton energy levels with even quantum numbers are dipole-forbidden states, that is optically forbidden [59]. However, the optical near-field interaction is finite for such forbidden energy state [26].

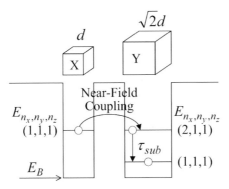

Fig. 3. *Upper*: Schematic drawings of closely located cubic CuCl QDs X and Y with the effective side lengths $(L-a_{\mathrm{B}})$ of d and $\sqrt{2}\,d$, respectively, where L and a_{B} are the side lengths of cubic QDs and the exciton Bohr radius, respectively. *Lower*: Their exciton energy levels. n_x, n_y, and n_z represent quantum numbers of an exciton. E_{B} is the exciton energy level in a bulk crystal

Figure 3 shows schematic drawings of the different-sized cubic CuCl QDs (X and Y) and confined-exciton Z_3 energy levels. Here, d and $\sqrt{2}\,d$ are the effective side lengths of cubic QDs X and Y, respectively. According to (1), the quantized exciton energy levels of $(1,1,1)$ in QD X and $(2,1,1)$ in QD Y resonate with each other. For this type of resonant condition, the coupling energy of the optical near-field interaction is given by the following Yukawa function [2, 26]:

$$V(r) = A\frac{\exp(-\mu \cdot r)}{r}\,. \tag{2}$$

Here, r is the separation between two QDs, A is the coupling coefficient, and the effective mass of the Yukawa function μ is given by

$$\mu = \frac{\sqrt{2E_{n_x,n_y,n_z}\left(E_{\mathrm{NaCl}} + E_{n_x,n_y,n_z}\right)}}{\hbar c}\,, \tag{3}$$

where E_{NaCl} is the exciton energy of an NaCl matrix. The value of the coupling coefficient A depends on each experimental condition; however, we can estimate it from the result of the previous work on the interaction between an Rb atom and the optical near-field probe-tip [26]. The value of A for 5-nm CuCl QDs is found to be more than 10 times larger than that for the Rb-atom case, since the coupling coefficient A is proportional to the oscillator strength and square of the photon energy [19, 26]. Assuming that the separation r between two QDs is equal to 10 nm, the coupling energy $V(r)$ is the order of 10^{-4} eV. This corresponds to a transition time of 40 ps, which is much shorter than the exciton lifetime of a few ns. In addition, inter sublevel transition τ_{sub} from higher exciton energy levels to the lowest one, as shown in Fig. 3, is generally less than a few ps and is much shorter than the transition time due to optical near-field coupling [63]. Therefore, most of the energy of the

exciton in a cubic CuCl QD with the side length of d transfer to the lowest
exciton energy level in the neighboring QDs with a side length of $\sqrt{2}\,d$ and
recombine radiatively in the lowest level.

2.2 Experimental Results and Discussions

We fabricated CuCl QDs embedded in an NaCl matrix by the Bridgman
method and successive annealing, and found the average size of the QDs to be
4.3 nm. The sample was cleaved just before the NFO spectroscopy experiment
in order to keep the sample surface clean. The cleaved surface of the sample
with 100-μm-thick sample was sufficiently flat for the experiment, i.e., its
roughness was less than 50 nm at least within a few μm squares. A 325-nm
He–Cd laser was used as a light source. A double-tapered fiber probe was
fabricated by chemical etching and a 150-nm gold coating was applied [64].
A 50-nm aperture was fabricated by the pounding method [65].

The curve in Fig. 4a shows a far-field luminescence spectrum of the sample
that was recorded with a probe-sample separation of 3 μm in the collection-
mode operation [2] of the cryogenic NFO microscope at 15 K. It represents
the collective luminescence intensity from several CuCl QCs, and is inhomogeneously broadened owing to the size distribution of the QDs. Figure 4b

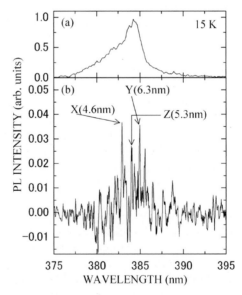

Fig. 4. (a) Far-field luminescence spectrum of a sample recorded with probe-sample
separation of 3 μm for the collection-mode operation at 15 K. (b) The differential
luminescence spectrum, which is the intensity difference between luminescence measured with the probe-sample separations of 3 μm and of less than 10 nm. X, Y, and
Z correspond to the wavelengths of lowest exciton state in cubic QDs with the side
lengths of 4.6, 6.3, and 5.3 nm, respectively

represents the differential spectrum, which is the intensity difference between luminescence measured with probe-sample separations of $3\,\mu m$ and of less than $10\,nm$. This curve consists of many fine structures. These appear as the contribution of the QDs near the apex of the probe because of the drastic increase in the measured luminescence intensity for a probe-sample separation less than $10\,nm$. The average density of the QDs is $10^{17}\,cm^{-3}$. Thus, the average separation between the QDs is less than $30\,nm$, estimated from the concentration of CuCl. Therefore, the spectral peaks in Fig. 4b, obtained by near-field measurement using the 50-nm aperture fiber probe, originate from several QDs of different size. Among these, the peaks X and Y correspond to the confined Z_3-exciton energy levels of quantum number $(1,1,1)$ for the cubic QDs with side lengths (L) of 4.6 nm and 6.3 nm, respectively. Their effective side lengths d are 3.9 and 5.6 nm, whose size ratio is close to $1 : \sqrt{2}$, and thus the $(1,1,1)$ and $(2,1,1)$ quantized exciton levels are resonant with each other to be responsible for energy transfer between the QDs.

Figure 5a and b shows the spatial distributions of the luminescence intensity, i.e., NFO microscope images, for peaks X and Y in Fig. 4b, respectively. The spatial resolution, which depends on the aperture diameter of the near-field probe, was smaller than 50 nm. These images clearly establish anti-correlation features in their intensity distributions, as manifested by the dark and bright regions surrounded by white broken curves. In order to confirm the anti-correlation feature more quantitatively, Fig. 6 shows the values of the cross-correlation coefficient C between the spatial distribution of the intensity of the luminescence emitted from the (n_x, n_y, n_z) level of exciton in a cubic QD with 6.3-nm side length and that from the $(1,1,1)$ level in a QD with a different side length L. They have been normalized to that of the auto-correlation coefficient of the luminescence intensity from the $(1,1,1)$ level in a 6.3-nm QD, which is identified by an arrow (1) in Fig. 6. To calculate the values of C, spatial Fourier transform was performed on the series of luminescence intensity values in the chain of pixels inside the region surrounded by the broken white curves in Fig. 5. The large negative value of C identified by an arrow (2) clearly shows the anti-correlation feature between Fig. 2a and b, i.e., between the $(2,1,1)$ level in a 6.3-nm QD and the $(1,1,1)$ level in a 4.6-nm QD.

This anti-correlation feature can be understood by noting that these spatial distributions in luminescence intensity represent not only the spatial distributions of the QDs, but also some kind of resonant interaction between the QDs. This interaction induces energy transfer from X dots ($L = 4.6\,nm$) to Y dots ($L = 6.3\,nm$). Interpreting this, most of 4.6-nm QDs located close to 6.3-nm QDs, cannot emit light, but instead transfer the energy to the 6.3-nm QDs. As a result, in the region containing embedded 6.3-nm QDs, the luminescence intensity in Fig. 5a from 4.6-nm QDs is low, while the corresponding position in Fig. 5b is high. As we mentioned above, it is reasonable to attribute the origin of the interaction to the near-field energy transfer. Besides, anti-correlation features appear for every couple of QDs with different sizes

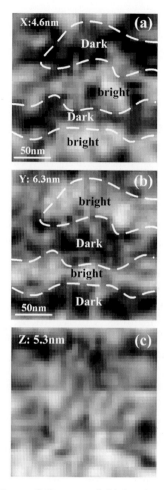

Fig. 5. Spatial distributions of the near-field luminescence intensity for peaks marked as X, Y, and Z in Fig. 4b, respectively

satisfying the resonant conditions of the confinement exciton energy levels. Therefore, we claimed that this is the first observation of energy transfer between QDs via the optical near field.

For reference, we note the dark area outside the broken curves in Fig. 5b. This occurs because there are very few 6.3-nm QDs. From the absorption spectrum of the sample, it is estimated that the population of 6.3-nm QDs is one-tenth the population of 4.6-nm QDs. As a result, the corresponding area in Fig. 5a is bright due to absence of the energy transfer.

On the other hand, the spatial distributions of the luminescence intensities from other QDs whose sizes do not satisfy the resonant condition given by (1) did not show any anti-correlation features. This is confirmed by comparing

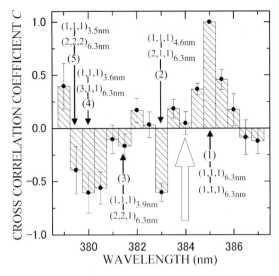

Fig. 6. Values of the cross-correlation coefficient C between the spatial distribution of the intensity of the luminescence emitted from the (n_x, n_y, n_z) level of exciton in a QD with 6.3-nm side length and that from the $(1, 1, 1)$ level in a QD with the different side length L. They have been normalized to that of the auto-correlation coefficient of the luminescence intensity from the $(1, 1, 1)$ level in a 6.3-nm QD, which is identified by an *arrow* (1). Other four *arrows* (2)–(5) represent the cross-correlation coefficient C between higher levels in a 6.3-nm QD and other sized QDs. They are between: (2) the $(2, 1, 1)$ level in a 6.3-nm QD and the $(1, 1, 1)$ level in a 3.9-nm QD, (3) the $(2, 2, 1)$ level in a 6.3-nm QD and the $(1, 1, 1)$ level in a 3.9-nm QD, (4) the $(3, 1, 1)$ level in a 6.3-nm QD and the $(1, 1, 1)$ level in a 3.6-nm QD, and (5) the $(2, 2, 2)$ level in a 6.3-nm QD and the $(1, 1, 1)$ level in a 3.5-nm QD. For reference, a *white arrow* represents the value of C between the $(2, 1, 1)$ level in a 6.3-nm QD and the nonresonant $(1, 1, 1)$ level in a 5.3-nm QD

Fig. 5a–c. Here, Fig. 5c shows the spatial distribution of the luminescence intensity of peak Z in Fig. 4b, which corresponds to cubic QDs with a side length of 5.3 nm. The white arrow in Fig. 6 indicates the relationship between Fig. 5b and c. The negligibly small value of C identified by this arrow proves the absence of the anti-correlation feature between the exciton energy levels in a 6.3-nm QD and the $(1, 1, 1)$ level in a 5.3-nm QD due to their nonresonant condition.

Furthermore, arrows (3)–(5) also represent clearly large negative values of C, which means the existence of the anti-correlation feature between higher levels in 6.3-nm QD and other sized QDs. They are: (3) the $(2, 2, 1)$ level in a 6.3-nm QD and the $(1, 1, 1)$ level in a 3.9-nm QD, (4) the $(3, 1, 1)$ level in a 6.3-nm QD and the $(1, 1, 1)$ level in a 3.6-nm QD, and (5) the $(2, 2, 2)$ level in a 6.3-nm QD and the $(1, 1, 1)$ level in a 3.5-nm QD. These anti-correlation features can also be explained by the resonant optical near-field energy transfer.

The features represented in this figure support our interpretation of the experimental results. The large anti-correlation coefficients C identified by arrows (4) and (5) in Fig. 6 are the evidence of multiple energy transfer: since the $(1,1,1)$ levels in 3.5- and 3.6-nm QDs resonate or nearly resonate to the $(2,1,1)$ level in a 4.6-nm QD, there is another route of energy transfer in addition to direct transfer from the 3.5- and 3.6-nm QDs to 6.3-nm QDs, i.e., the transfer via the 4.6-nm QDs. We consider that such multiple energy transfers increase the value of C identified by arrows (4) and (5) in Fig. 6.

The anti-correlation features appear for every pair of QDs if the resonant conditions of the confinement exciton energy levels are satisfied. Figure 7 shows two-dimensional (2D) plots of cross-correlation coefficient C between two QDs. White solid lines indicate the QDs pairs with resonant energy levels of $(2,1,1)$–$(1,1,1)$, $(2,2,1)$–$(1,1,1)$, $(3,1,1)$–$(1,1,1)$, and $(2,2,2)$–$(1,1,1)$, respectively. Large negative appears along the solid lines due to optical near-field energy transfer between QDs.

Here we demonstrate the evidence of the near-field energy transfer between QDs using NFO spectroscopic microscopy. This phenomenon can provide a variety of applications, such as an all-optical near-field device, as shown in the later sections.

2.3 Anti-Parallel Dipole Coupling of Quantum Dots

For a coupled QD system, the carrier lifetime is expected to differ from that of an isolated QD. Figure 8 shows schematic drawings of the typical states of coupled QDs which are much smaller than the wavelength of light. When their

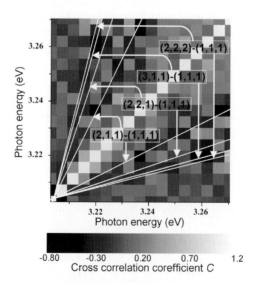

Fig. 7. 2D plots of cross-correlation coefficient C between two cubic QDs. *White solid lines* show the QDs pairs with resonant energy levels of $(2,1,1)$–$(1,1,1)$, $(2,2,1)$–$(1,1,1)$, $(3,1,1)$–$(1,1,1)$, and $(2,2,2)$–$(1,1,1)$, respectively

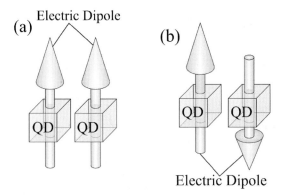

Fig. 8. Schematic drawing of a QD pair and its electric dipoles. (**a**) The electric dipoles are parallel to each other, i.e., dipole-allowed and the superradiant state. (**b**) The dipoles are anti-parallel, i.e., the dipole-forbidden state

electric dipoles are parallel, the coupled energy state shows dipole-allowed and the carrier lifetime decreases due to the increase in the total oscillator strength, i.e., Dicke's superradiance [10], as shown in Fig. 8a. Conversely, when their electric dipoles are anti-parallel, the coupled energy state shows dipole forbidden and their carrier lifetime increases, because the total oscillator strength decreases, as shown in Fig. 8b. The electric lines of force are quite different each other. It is interesting that the coupled QDs via optical near field shows whether dipole allowed or forbidden states.

Here, we used the cubic CuCl QDs with the average side length $L = 4.2$ nm. A 325-nm CW He–Cd laser and 385-nm SHG of CW and mode-locked Ti-sapphire lasers (repetition rate: 80 MHz) were used as the light sources. The duration of the transform-limited pulse of the mode-locked laser was set at 10 ps. A double-tapered fiber probe with a 150-nm aluminum coating and a 40-nm diameter aperture was used. After the QD pairs in the inhomogeneous size-dispersed sample were found using an optical near-field microscope, the temporal evolution of the photoluminescence (PL) pump–probe signal was detected using the time correlation single photon counting method with a 15-ps time resolution.

Figure 9a and b shows the spatial distribution of the luminescence intensity and the near-field PL spectrum of the sample from a 6.3-nm CuCl QD at 15 K, respectively, with the 325-nm CW probe light only, which excited the band-to-band transition in the sample. The inset in Fig. 9a shows the energy transfer between the observed QDs, i.e., from 4.6- to 6.3-nm QDs, where τ_i, τ_{sub}, and τ_{ex} are the energy transfer time, inter-sublevel transition time, and exciton lifetime, respectively. As we mentioned in Sect. 2.2, there was resonance between the quantized exciton energy level of quantum number $(1, 1, 1)$ in the 4.6-nm QDs and the quantized exciton energy level of quantum number $(2, 1, 1)$ in the 6.3-nm QDs. The luminescence of a 4.6-nm QD decreases due to competitive inhibition and that of a 6.3-nm QDs increases due to the supply

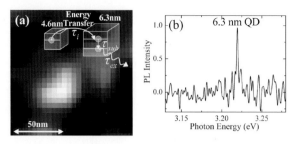

Fig. 9. (a) The spatial distribution of the luminescence intensity from the 6.3-nm QD with the 325-nm CW probe light only. The *inset* shows the observed QD pair and the energy flow. (b) The near-field PL spectrum at the position where the strong luminescence was observed in (a)

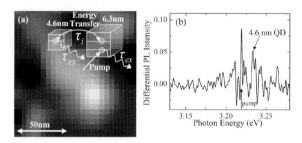

Fig. 10. (a) The spatial distribution of the luminescence intensity from the 4.6-nm QD with the 325-nm CW probe light and the 385-nm 10-ps pump pulse. Here we used the narrow band-pass filter (FWHM: 8 meV), whose optical density in the stop-band is more than 6. The *inset* shows the observed QD pair and the energy flow. (b) The near-field PL spectrum at the position where the strong luminescence was observed in (a)

of the excitation energy from the neighboring 4.6-nm QD. As a result, the PL signal from the 6.3-nm QD was observed as the spectral peak, as shown in Fig. 9b.

Figure 10a and b shows the spatial distribution of the luminescence intensity from the 4.6-nm QD and the differential PL spectrum and at 15 K, respectively, taken with the 325-nm CW probe light and the 385-nm 10-ps pump pulse. Here, the differential PL spectral intensity is defined as (the PL spectrum with pump and probe light) − (the PL spectrum with the probe light only) − (the PL spectrum with the pump light only). The upward pointing arrow in Fig. 10b shows the photon energy of the pump pulse tuned to the $(1,1,1)$ exciton energy level in the 6.3-nm QD. The inset in Fig. 10a shows the energy transfer between the QDs when the pump pulse excites the 6.3-nm QD. In this case, because the exciton energy in the 4.6-nm QD cannot be transferred to the $(1,1,1)$ exciton energy level in the 6.3-nm QD due to the state filling effect, the exciton energy flows back-and-forth between the $(1, 1, 1)$ exciton energy level in the 4.6-nm QD and $(2, 1, 1)$ exciton energy level in the 6.3-nm QD [17, 18], and some excitons recombine in the 4.6-nm QD. Therefore, the PL signal from the 4.6-nm QD was detected as the spectral peak indicated by the arrow in Fig. 10b. The temporal evolution of this PL signal strongly depends on τ_i and τ_{ex} of the coupled QD system [17, 18].

We found the coupled QDs, i.e., the QD pair of 4.6- and 6.3-nm QDs, and measured the temporal evolution of the differential PL signal. Figure 11a shows the temporal evolution of the differential PL peak intensity from 4.6-nm QDs of the different QD pairs. The open squares (P1), circles (P2), and triangles (P3) correspond to the experimental results observed for three different 4.6- and 6.3-nm QDs pairs. The solid, broken, and dotted lines (at the rise and decay of the time evolution of the signals) are fitted to the experimental values using the simple rate equation using respective τ_i and τ_{ex} for the 4.6-nm QD and the 6.3-nm QD as fitting parameters. In Fig. 11a, the exciton

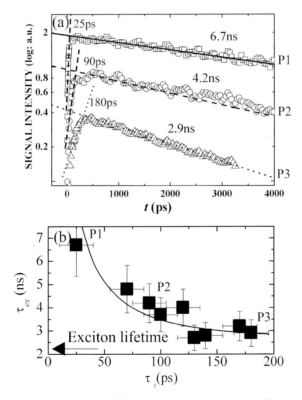

Fig. 11. Time evolutions of the PL peak signal intensity in Fig. 10a observed at different positions in the sample, i.e., different QD pairs, (P1, P2, and P3). (b) Relationship between the τ_i and τ_{ex} obtained from the fitting to the temporal evolution in (a)

population in the 6.3-nm QD is increased due to the pump pulse at $t = 0$. The exciton population in the 4.6-nm QD is also increased due to the prohibited energy transfer to the 6.3-nm QD with the filling effect.

The solid squares in Fig. 11b are the experimental results for the relation between τ_{ex} and τ_i of the PL from the 4.6-nm QD for several QD pairs including P1, P2, and P3, where the τ_{ex} and τ_i were obtained from the fitting. The τ_{ex} exceeds the exciton lifetime of the isolated 6.3-nm QD measured experimentally, and increases as the τ_i falls. The optical near-field interaction is given by τ_i^{-1} [17,18]. Therefore, the experimental result in Fig. 11b means that the exciton lifetime in the coupled QDs increases with the increase in optical near-field interaction. This increase in the exciton lifetime due to the optical near-field interaction can be understood using the feature of the anti-parallel dipole–dipole coupling of an optical near field.

Finally, we discuss the origins of the anti-parallel coupling features of the optical near-field interaction between QDs. In the experiment, we detected the PL signal from QDs, which means that only the transverse exciton was detected, because the longitudinal exciton is optically forbidden and its dispersion differs from that of the transverse exciton, i.e., it has a different energy in the QDs. Since, the direction of the electric dipole in the transverse exciton is perpendicular to the direction of propagation, the dipole never becomes aligned with the direction of propagation after the energy is transferred to the neighboring QD. Although there are two possible eigenstates of the mutual arrangements of the dipoles in excitons, i.e., parallel and anti-parallel, as shown in Fig. 8a and b, the occurrence probability of the anti-parallel state exceeds that of the parallel state because the total energy of the system for the anti-parallel state is lower than that for the parallel state. This anti-parallel feature of the optical near-field coupled QDs reduces the recombination of excitons. Consequently, the exciton lifetime increases with the optical near-field interaction. These features are of interest physically and are applicable to photonic devices, such as optical nanometric sources, long phosphorescence devices, and optical battery cells.

3 Demonstration of Nanophotonic Switch

3.1 Operation Principle of Nanophotonic Switch

First, we explain about a nanophotonic switching operation using CuCl QDs embedded in an NaCl matrix [21, 22]. When closely spaced QDs with quantized energy levels resonate with each other, near-field energy transfer occurs between them, even if the transfer is optically forbidden, as we mentioned in Sect. 2. Figure 12a and b explains the OFF and ON states of the proposed nanophotonic switch. Three cubic QDs, QD_{in}, QD_{out}, and $QD_{control}$, are used as the input, output, and control ports of the switch, respectively. Assuming an effective size ratio of $1{:}\sqrt{2}{:}2$, quantized energy levels $(1, 1, 1)$ in QD_{in}, $(2, 1, 1)$ in QD_{out}, and $(2, 2, 2)$ in $QD_{control}$ resonate with each other.

(a) OFF state

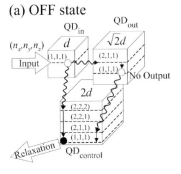

(b) ON state

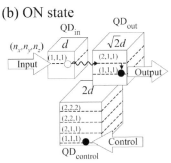

Fig. 12. Schematic explanation of a nanophotonic switch. The closely located cubic CuCl QDs (QD$_{\text{in}}$, QD$_{\text{out}}$, and QD$_{\text{control}}$) with the effective side lengths of d ($= L - a_{\text{B}}$), $\sqrt{2}\,d$, and $2d$ respectively, where L and a_{B} are the side lengths of the cubic QDs and the exciton Bohr radius, respectively. n_x, n_y, and n_z represent quantum numbers of confined excitons. (a) and (b) show, respectively, the OFF and ON states of the switch using three QDs

Furthermore, energy levels $(1, 1, 1)$ in QD$_{\text{out}}$ and $(2, 1, 1)$ in QD$_{\text{control}}$ also resonate. In the OFF operation (Fig. 12a), almost all of the exciton energy in QD$_{\text{in}}$ is transferred to the $(1, 1, 1)$ level in the neighboring QD$_{\text{out}}$, and finally, to the $(1, 1, 1)$ level in QD$_{\text{control}}$. Thus, the input energy escapes to QD$_{\text{control}}$, and consequently no optical output signals are generated from QD$_{\text{out}}$. In the ON state (Fig. 12b), by contrast, the escape routes to QD$_{\text{control}}$ are blocked by the excitation of QD$_{\text{control}}$, due to state filling in QD$_{\text{control}}$ by applying the control signal; thus, the input energy is transferred to QD$_{\text{out}}$ and an optical output signal is generated. The coupling energy $V(r)$ of the optical near-field interaction is given by the Yukawa function (2) [26], where r is the separation between the two QDs, A is the coupling coefficient, and μ is proportional to the effective mass of an optical near-field photon. Assuming that the separation r between the two QDs is 10 nm, the coupling energy, $V(r)$, is estimated to be on the order of 10^{-4} eV, which corresponds to a transfer time of 20 ps. Since the near-field interaction is a local interaction, as indicated by $V(r)$, it is applicable to the operation of a nanophotonic switch and relevant devices in nanophotonic integrated circuits.

3.2 Experimental Results and Discussions

In the experiment, we used CuCl QDs embedded in an NaCl matrix, as we mentioned in Sect. 2. The used double-tapered UV fiber probe was fabricated using chemical etching and coated with 150-nm-thick Al film. An aperture less than 50-nm in diameter was formed by the pounding method [65].

To realize the switching operation, the fiber probe was carefully scanned to search for a trio of QDs that satisfied an effective size ratio of $1:\sqrt{2}:2$ for the switching operation. Since the homogeneous linewidth of a CuCl QD increases with temperature [66, 67], the allowance in the resonantable size ratio is 10% at 15 K. The separation of the QDs must be less than 30 nm for the proposed switching operation, because the energy transfer time increases with separation; however, it must be shorter than the exciton lifetime. It is estimated that at least one trio of QDs exists in a $2\,\mu m \times 2\,\mu m$ scan area that satisfies these conditions. To demonstrate the switching operation, we had to find a set of QD-trio, as shown in Fig. 12, in the sample.

Near-field PL pump–probe spectroscopy to find the QD trio was carried out by fixing the fiber probe under the three excitation conditions: (1) only with a pump laser only ($\lambda = 385$ nm), (2) only with a probe laser only ($\lambda = 325$ nm), and (3) with pump and probe lasers. The wavelength of pump laser was tuned to the $(1, 1, 1)$ exciton energy level in a 6.3-nm CuCL QD. To clarify the PL difference with and without pump, the differential PL intensity (PL_{diff}) was defined as $PL_{diff} = PL_{pump\&probe} - PL_{pump} - PL_{probe}$, where PL_{pump}, PL_{probe}, and $PL_{pump\&probe}$ are the PL intensities measured under conditions (1)–(3), respectively. Figure 13 shows the PL_{diff} spectrum

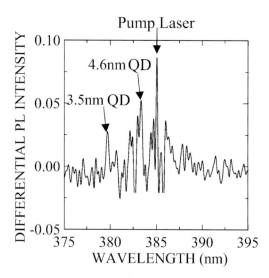

Fig. 13. Near-field differential PL spectra measured at position where the QD trio exists acting as the nanophotonic switch

obtained at position where the QD trio exists. In Fig. 13, two satellite peaks appear at the positions of the $(1, 1, 1)$ levels in the 4.6- and 3.5-nm QDs. The appearance of the satellite peaks means that the switching system proposed in Fig. 12 was constructed in the area under the probe. In other words, a trio of cubic QDs with sizes of 3.5, 4.6, and 6.3 nm. Since their effective sizes $L - a_B$ were 2.8, 3.9, and 5.6 nm (a_B: exciton Bohr radius), respectively, the size ratio was close to $1 : \sqrt{2} : 2$ and they could be used as QD_{in}, QD_{out}, and $QD_{control}$, respectively. The pumping to the 6.3-nm QD blocks the energy transfer from the 3.5- and 4.6-nm QDs to the 6.3-nm QD due to the state filling of the 6.3-nm QD, and the 3.5- and 4.6-nm QDs emit light that results in the satellite peaks in Fig. 13. Thus, we found the QD-trio acting as the nanophotonic switch.

The PL peak from the 4.6-nm QD corresponds to the output signal in Fig. 12b. The PL_{diff} intensity from the 4.6-nm QD, was 0.05 ± 0.02 times the PL intensity from the 6.3-nm QD, which obtained only with the probe laser. This value is quite reasonable considering the pumping power density of 10^3 W cm^{-2}, because the probability density of excitons in a 6.3-nm QD is 0.1 ± 0.05 [66], which is close to the PL_{diff} intensity from the 4.6-nm QD. This result indicates that the internal quantum efficiency of the switching system is close to 1.

In the experiment of the switching operation, 379.5- and 385-nm SHG Ti:sapphire lasers, which were tuned to the $(1, 1, 1)$ exciton energy levels of QD_{in} and $QD_{control}$, respectively, were used as the input and the control light sources. The output signal was collected by the fiber probe, and its intensity was measured by a cooled microchannel plate after passing through three interference filters of 1-nm bandwidth tuned to the $(1, 1, 1)$ exciton energy level in QD_{out} at 383 nm.

Figure 14a and b shows the spatial distribution of the output signal intensity in the OFF state, i.e., with an input signal only, and the ON state, i.e., with input and control signals, using near-field spectroscopy at 15 K. The insets in Fig. 14 are schematic drawings of the existing QD trio used for the switching, which was confirmed by the near-field luminescence spectrum. Here the separation of the QDs by less than 20 nm (drawings in Fig. 14) was theoretically estimated from time-resolved measurements, as explained in later in this chapter. In the OFF state, no output signal was observed, because the energy of the input signal was transferred to $QD_{control}$ and swept out as luminescence at 385 nm. To quench the output signal in the OFF state, which is generated by exciton accumulation in $QD_{control}$, the input signal density in $QD_{control}$ was regulated to less than 0.1 excitons. In the ON state, we obtained a clear output signal in the broken circle. The output signal was proportional to the intensity of the control signal, where its density in $QD_{control}$ was 0.01–0.1 excitons.

Next, the dynamic properties of the nanophotonic switch were evaluated by using the time-correlation single-photon counting method. As a pulse-control light source, the 385-nm SHG of a mode-locked Ti-sapphire laser was used. The repetition rate of the laser was 80 MHz. To avoid cross-talk of the input

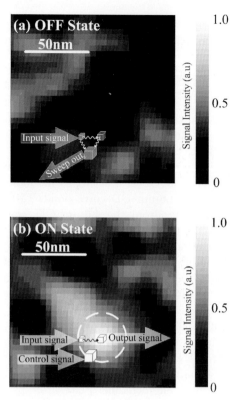

Fig. 14. Spatial distribution of the output signal from the nanophotonic switch measured by a near-field microscope, in the OFF (**a**) and ON (**b**) states

and control signals, which originates from spectral broadening for the pulse duration, the pulse duration of the mode-locked laser was set to be 10 ps. The time resolution of the experiment was 15 ps. Figure 15 shows the time evolution of the control pulse signal (upper part) applicable to $QD_{control}$ and the output signal (lower part) from QD_{out}. The output signal rises synchronously within less than 100 ps, with the control pulse, which agrees with the theoretically expected result based on the Yukawa model [26]. As this signal rise time is determined by the energy transfer time between the QDs, the separation between the QDs can be estimated from the rise time as being less than 20 nm; the rise time can be shortened to a few ps by decreasing the separation of the QDs. Since the decay time of the output signal is limited by the exciton lifetime, this nanophotonic switch is able to operate at a few hundred MHz, and it is anticipated that the operation frequency can be increased to several GHz by means of exciton quenching using a plasmon coupling [68]. The ON–OFF ratio was about 10, which is sufficient for use as an all-optical switch, and can be increased by a saturable absorber and electric field enhancement of the surface plasmon [69].

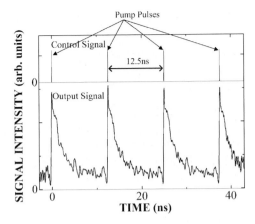

Fig. 15. Time evolution of the control pulse signal (*upper* part) and the output signal (*lower* part) from the nanophotonic switch at the *broken circle* in Fig. 14b. The duration and repetition rate of the control pulse were 10 ps and 80 MHz, respectively

Table 1. Figure of merits of optical switches

classification	size:V	switching time	switching energy	ON–OFF contrast	figure of merit
MEMS	$(n\lambda)^3$	1 µs	10^{-18} J	10^4	10^{-5}
Mach-Zehnder	$(n\lambda)^3$	10 ps	10^{-18} J	10^2	10^{-2}
nonresonat, $\chi^{(3)}$	$(n\lambda)^3$	10 fs	10^6 photons	10^3	10^{-3}
resonat, $\chi^{(3)}$	$(n\lambda)^3$	1 ns	10^3 photons	10^4	10^{-4}
sublevel	$(n\lambda)^3$	100 fs	10^3 photons	10^3	10^{-1}
nanophotonic	$(\lambda/10)^3$	∼100 ps	1 photon	10	1∼

The advantages of this nanophotonic switch are its small size and high-density integration capability based on the locality of the optical near field. The figure of merit (FOM) of an optical switch should be more important than the switching speed. Here, we have defined the FOM as $F = C/(V \times t_{sw} \times P_{sw})$, where C, V, t_{sw}, and P_{sw} are the ON–OFF ratio, the volume of the switch, the switching time, and the switching energy, respectively. Table 1 shows the comparison between the nanophotonic switch and the conventional photonic switch. The FOM of our demonstrated switch is 10–100 times higher than that of conventional photonic switches, because its volume and switching energy are 10^{-5} times and 10^{-3} times smaller, respectively. By conventional fabrication method, it is difficult to fabricate the nanophotonic devices, such as the QD trio functioning as the optical switch, because it is required that the designed sized QDs are placed at designed positions which are closely spaced. Therefore, we also research materials, fabrication methods, and systems, as we review later sections, for the realization of actual nanophotonic devices, as we mention in the later section.

4 Optical Nanofountain to Concentrate Optical Energy

4.1 Operation of Optical Nanofountain

The optical nanofountain is operated using this energy transfer, as shown in Fig. 16a, together with the energy transfer between QDs via an optical near-field interaction [16–18]. When closely spaced QDs with quantized energy levels resonate with each other, near-field energy transfer occurs between them. Assuming that an effective size ratio between closely located cubic QD-A and QD-B is $1:\sqrt{2}$, the quantified energy levels $(1,1,1)$ in QD-A and $(2,1,1)$ in QD-B resonate with each other, so that almost all of the excitation energy in QD-A is transferred to the $(1,1,1)$ level in QD-B via near-field energy transfer and successive inter-sublevel relaxation [63]. This unidirectional energy transfer from smaller to larger QDs concentrates the optical energy in a nanometric region in a biomimetic manner (see Sect. 1.3). When different sized QDs with resonant energy sublevels are distributed as shown in Fig. 16b, energy transfer occurs via the optical near field, as illustrated by the arrows. Light incident to the QD array is ultimately concentrated in the largest QD. The size of the area of light concentration corresponds to that of the QD. Therefore, this device realizes nanometric optical concentration. Since the mechanism

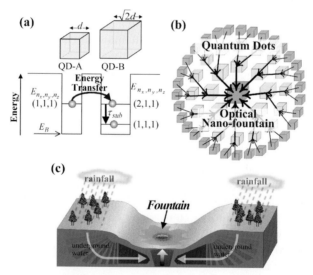

Fig. 16. (a) Schematic explanation of the energy transfer between QDs via an optical near-field interaction. E_{nx}, E_{ny}, E_{nz} $(n_x, n_y, n_z) = (1,1,1)$ or $(2,1,1)$ is the quantum number representing the excitonic energy level in a QD. (b) Schematic explanation of the optical nanofountain and unidirectional energy transfer. (c) Schematic drawing of a fountain in a basin

of the optical nanofountain is similar to that of the light-trapping system in photosynthetic bacteria, the operation of the optical nanofountain is a biomimetic action. The device proposed here is called an *optical nanofountain* because light spurts from the largest QD after it is concentrated by stepwise energy transfer from smaller neighboring QDs. In action, the device looks like a fountain in a basin, as shown schematically in Fig. 16c. From previous experimental tests of nanophotonic switch operation, as we mentioned in Sect. 3, it is expected that the concentration efficiency of this device will be close to 1 because there are no other possible relaxation paths in the nanometric system.

To demonstrate the operation of an optical nanofountain, we used CuCl cubic QDs embedded in an NaCl matrix with an average QD size of 4.2 nm and an average separation of less than 20 nm. Although, the QDs have an inhomogeneous size distribution and are randomly arranged in the matrix, the operation can be confirmed if an appropriate QD group is found using nanometric resolution near-field spectrometer. For the operation, we maintained a sample at the optimum temperature T (40 K). At $T < 40$ K, the resonant condition becomes tight due to narrowing of the homogeneous linewidth of the quantized energy sublevels, while at $T > 40$ K, the unidirectional energy transfer is obstructed by the thermal activation of excitons in the QDs. A 325-nm He–Cd laser was used as the excitation light source. A double-tapered UV fiber probe with an aperture 20 nm in diameter was fabricated using chemical etching and coated with a 150-nm-thick Al film to ensure sufficiently high detection sensitivity [64].

4.2 Experimental Results and Discussions

Figure 17a shows a typical near-field luminescence spectrum of the sample in the correction-mode operation [2]. It is inhomogeneously broadened due to the quantum size effect and the size distribution of the QDs. We have never observed luminescence of the exciton molecules due to the low excitation density of less than $1\,W\,cm^{-2}$. The spectral curve includes several fine peaks, which are the luminescence that comes from different sized QDs. Since we can obtain the size-selective QD position from the spatial distribution of the luminescence peak intensity, the 2D scanning measurement of the luminescence intensity collected by the photon energy allows us to search for QDs acting as optical nanofountains. At 40 K, it is not so difficult to find the QD array acting as optical nanofountain. We found about one optical nanofountain in $5\,\mu m \times 5\,\mu m$ region on the sample surface experimentally.

Figure 17b shows the typically spatial distribution of the luminescence emitted from QDs that operates well as an optical nanofountain. Here, the collected luminescence photon energy, E_p, was $3.215\,eV \leq E_p \leq 3.350\,eV$, which corresponded to the luminescence from QDs of size $2.5\,nm \leq L \leq 10\,nm$. The bright spot inside the broken circle corresponds to a spurt from an optical nanofountain, i.e., the focal spot of the nanometric optical condensing device.

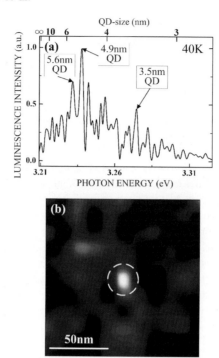

Fig. 17. (a) Near-field luminescence spectrum of CuCl QDs at 40 K. The relationship between the photon energy of luminescence and the size of the QDs is shown above and below the horizontal axes. (b) Spatial distribution of the luminescence intensity in an optical nanofountain. The bright spot surrounded by a broken circle is the focal spot

The diameter of the focal spot was less than 20 nm, which was limited by the spatial resolution of the near-field spectrometer. From the Rayleigh criterion (i.e., resolution $= 0.61 \cdot \lambda/\text{NA}$) [70], we obtained its numerical aperture (NA) of 12 for $\lambda = 385$ nm.

To demonstrate the detailed operating mechanism of the optical nanofountain, we show the size-selective luminescence intensity distribution, i.e., by photon energy, in Fig. 18a–c. The broken circles and the areas scanned by the probe are equivalent to that in Fig. 17b. The luminescence intensity distribution is displayed using a grey scale, whose normalized scales are (a) 0–0.6, (b) 0–0.2, (c) 0–0.1, and (d) 0–1. Cubes represent QDs whose positions were estimated from the luminescence intensity distribution. In Fig. 18a, a single QD of 6 nm $\leq L \leq$ 10 nm was observed at the focal position. In Fig. 18b and c, the observed QDs are 4 nm $\leq L \leq$ 6 nm and 2.5 nm $\leq L \leq$ 4 nm, respectively, and are located around the broken circles. Figure 18d shows the total luminescence intensity distribution obtained as the integral of Fig. 18a–c. The bright spot in this figure agrees with the position of the largest QD in Fig. 18a and

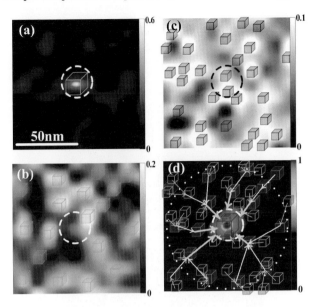

Fig. 18. Spatial distribution of the luminescence intensity of CuCl QDs of (**a**) 6 nm $\leq L \leq$ 10 nm (3.215 eV $\leq E_p \leq$ 3.227 eV), (**b**) 4 nm $\leq L \leq$ 6 nm (3.227 eV $\leq E_p \leq$ 3.254 eV), (**c**) 2.5 nm $\leq L \leq$ 4 nm (3.271 eV $\leq E_p \leq$ 3.350 eV), and (**d**) the total for 2.5 nm $\leq L \leq$ 10 nm (3.215 eV $\leq E_p \leq$ 3.350 eV) for the same area as in Fig. 17b. The *cubes* represent the positions estimated from the luminescence intensity distribution

the smaller QDs are distributed around it. This means that the optical energy is concentrated to the largest QD. The luminescence intensity at the bright spot is more than five times greater than that from a single isolated QD of $L = 10$ nm. While the luminescence intensities of the smaller surrounding QDs are lower than those of the isolated QDs. This indicates that optical energy is transferred from smaller to larger QDs and is concentrated in the largest QD, as shown by arrows in Fig. 18d. This device can also be used as a frequency selector, based on the resonant frequency of the QDs, which can be applied to frequency domain measurements, multiple optical memories, multiple optical signal processing, frequency division multiplexing, and so on. The application of the optical nanofountain are discussed in the other section by Naruse et al.

5 Laterally Coupled GaN/AlN Quantum Dots

The surface morphology of GaN nanostructures was studied using scanning electron (SEM) and atomic force microscopy (AFM). Shown in Fig. 19a is a topographical map of an area of 1 μm × 1 μm measured using AFM, which

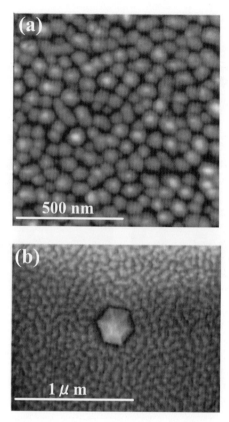

Fig. 19. (a) AFM image showing the surface of GaN dots covered with 2 nm AlN cap layer. (b) SEM images showing the surface morphology and self-assembly of a hexagonal pyramid shape GaN structure with 300 nm in diameter

exhibits a honeycomb feature at the surface due to modulation of the AlN cap layer by the underlying GaN QDs. Due to the thin cap layer, a high density $(3 \times 10^{10} \, \text{dots cm}^{-2})$ of the QDs as well as a strong inhomogeneity of their lateral dimensions, ranging from 30 up to 50 nm, is clearly evidenced. The height of these capped QDs range from 7 to 10 nm. This nonuniform surface topology induces inhomogeneous broadening in the far-field emission spectrum due to lateral coupling. It was shown by Widmann et al. [34] that the QD size varies significantly depending on whether the QDs are allowed to evolve under vacuum before covering with AlN, or not, as a result of a ripening mechanism. This variation in size can lead to a large variation in the piezoelectric effect in the self-assembled GaN layers. Our experiments indicate that ripening leads to reduced footprint and increased height for a larger aspect ratio, as the dots are not spherical [37].

The modulations at the surface are also observed in the high magnification SEM images (Fig. 19b). The SEM spatial patterns of the capped GaN QDs showing island-like features can be correlated with the sample morphology as measured by AFM. A large hexagonal GaN pyramid is self-assembled on the AlN cap on the surface of the GaN QD layers with a radius of curvature no more than 300 nm. The faces of the pyramids are the $\{10\bar{1}1\}$ planes as evidenced by the angle between the inclined edge and the base of the pyramid. The measured angle of around 58–60° is in good agreement with the calculated angle of 58.4° using the GaN lattice parameters of $c = 5.185$ Å and $a = 3.189$ Å. The formation of the pyramids indicates that the $\{10\bar{1}1\}$ surfaces are self-assembled preferentially compared to the $\{000\bar{1}\}$ surface. Thus, it can be inferred that $\{10\bar{1}1\}$ surfaces have the lowest surface potential with respect to the self-assembly process. The tip of the pyramid is very sharp with a diameter measured to be less than 20 nm.

The optical emission properties were investigated by studying the photoluminescence (PL) characteristics in the far-field and near-field limit. Figure 20 shows the time-integrated far-field PL spectrum of QDs at room temperature, measured using a frequency tripled Ti:sapphire laser delivering pulses of 10 ps duration at 267 nm (photon energy 4.655 eV). The peak emission energy was close to 3.67 eV, with a broad linewidth of 250 meV arising due to the inhomogeneous strain and also the lateral and vertical coupling amongst the QDs in the various layers. The PL peaks from the QD layers are shifted to a higher energy as compared to the underlying bulk GaN for the wurtzite phase (band gap energy $E_g = 3.45$ eV). The inhomogeneously broadened PL line shape can be attributed to the emission from optically pumped carriers thermalized in the statistically distributed ground states of the probed QD, which vary in energy because of small variations in size, composition, and strain. The inset shows the temperature dependence of the PL intensity and emission linewidth. The relatively temperature-insensitive PL emission below 125 K occurs as the radiative decay of excited carriers dominates the recombination process [38].

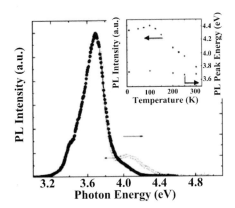

Fig. 20. Far-field PL spectrum of GaN QDs with temperature dependence of PL intensity and PL emission energy shown in the inset

However, above 125 K the PL intensity decreases more severely with increasing temperature due to increase in the nonradiative recombination. The relatively small change in thermally induced PL peak energy shift $(0.168\,\text{meV}\,\text{K}^{-1})$ is due to strong carrier confinement in the QDs with the redshift at higher temperatures likely due to a reduction of the exciton Bohr radius that makes the excitons less polar [35]. The PL excitation spectrum measured using a Xe lamp shows absorption from the GaN nanostructures from higher energies. A large Stark shift exceeding 400 meV is observed to the built-in strain in the QD layers.

Compared to bulk or GaN/AlN quantum well (QW), a larger PL efficiency has been observed for this QD system despite a relatively shorter radiative lifetime of 500 ps [38]. The role of inhomogeneity in the far-field PL spectra due to spatial QD distribution has been investigated via near-field PL spectroscopy. We have used a near-field microscope operating in the illumination mode at 10 K for measuring the spatially and spectrally resolved PL spectra. A tapered, metal-coated optical fiber having a nominal apical aperture of 30 nm was exploited as the nanosource through which the sample was irradiated with UV light (325 nm delivered by an He–Cd CW laser). Figure 17a–c shows monochromatic PL images within a 450 nm × 450 nm area, in which the detection wavelengths are 343 ± 1, 345 ± 1.5, and 355 ± 1 nm, respectively. The near-field PL in Fig. 21a and c, which originates from a much smaller number of QDs compared to dot density obtained from surface features in Fig. 19, consists of a number of sharp spectral features of similar amplitude with full width at half maximum (FWHM) ranging from 500 µeV to 2 meV. We observe that the bright areas in Fig. 21a and c are larger than the dark ones. It may be that the honeycomb-like QD features observed at the surface are not entirely optically active and larger islands or smaller QDs presumably act as nonradiative recombination centers. The nucleation of relatively larger dots emitting at lower energies, i.e., ∼3.49 eV (Fig. 21c) is more prevalent compared to the smaller dots emitting at higher energies ∼3.62 eV (Fig. 21a). The brighter regions A, B, C, E, F in Fig. 21a and c are an indication of strong confinement and a correlation in the vertical direction.

As the optically active part of our sample consists of 40 QD or QW planes, several layers with varying QD spatial distribution are excited simultaneously. So even if the near-field probe is located above a region of the first QD layer containing large dots, luminescence at high energy still originated from the underlying QD planes. This suggests that, while the intense background signal is due to the luminescence of a large part of the active region that cannot be spatially resolved, the localized modulations are due only to the morphology of the dots located on the outermost layer, which can be stronger in the presence of vertical correlation.

To gain insight into the origin of the light emission and the influence of spatial variation of GaN QDs and QWs on the intensity, we performed cross-sectional transmission electron microscopy (TEM). Samples were processed in a dual-beam SEM/FIB using a Ga ion beam accelerating voltage of 5 kV. A near vertical correlation of the GaN dots ∼30 nm in width is observed from

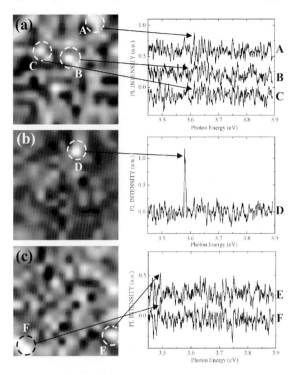

Fig. 21. Near-field luminescence spectra from GaN QD. (a) Spatial and spectrally resolved PL measured at 342–344 nm. (b) Spatial and spectrally resolved PL measured at 344–347 nm. (c) Spatial and spectrally resolved PL measured at 354–356 nm

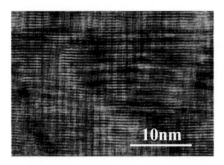

Fig. 22. HRTEM image showing {1-1-1-0} cross-section of stacked layers GaN dots

STEM-HAADF image, with some dot assemblies being correlated at an angle slightly off vertical. It is also observed that the width of these dots and their period correspond to the surface texture observed in AFM and SEM images (Fig. 19). An HRTEM image shown in Fig. 22 illustrates that 1.1- to 2-nm-high GaN QD-like clusters are embedded in GaN/AlN QW-like structures.

It is reasonable to assign the high energy PL spectrum (Fig. 20) to the superposition of blueshifted near-band-gap exciton-emissions arising from clusters of dots with size smaller than the exciton Bohr radius for GaN ($a_B \sim 3$ nm), at least in the growth direction (3 nm). The distribution of the dots in the vertically stacked layers also explains the background emission from spatially unresolved underlying QD and QW layers. The strong room-temperature PL is due to the vertical correlation of the dots, while the lateral coupling at the surface and underlying layers results in nonradiative recombination. Contrast observed in the near-field images (Fig. 21) may be due to stronger emission from dot clusters correlated more closely to the vertical direction, as opposed to dot clusters correlated off-axis observed in Fig. 22.

An intense emission is observed from a 20-nm-diam area D at an intermediate energy regime 3.59 eV (345 nm), with a small background emission (Fig. 21b), implying that the source of this strong PL is significantly different from the emission of larger QDs or smaller QDs shown in Fig. 21a and c. The emission at 3.39 eV is particularly strong in the vicinity of the hexagonal pyramid structure shown in Fig. 19b. The emission is likely due to the localization of excitons in GaN QD formed at the tip of the hexagonal pyramid. The strong room-temperature PL is due to the vertical correlation of the dots, while the lateral coupling at the surface and underlying layers results in nonradiative recombination resulting in reduced PL emission at higher temperatures.

GaN QDs is one of the promising matrial for nanophotonic device acting at room temperature. Other matrial for nanophotonic device is reviewed in other chapter by Yatsui et al.

6 Nonadiabatic Nanofabrication Using Optical Near Field

In this section, we review the nanofabrication technique using the optical near field, and discuss the unique nonadiabatic photochemical reaction using exciton–phonon polariton model.

6.1 Nonadiabatic Near-Field Optical CVD

Figure 23 shows the experimental setup for an NFO-CVD [42, 46]. Ultrahigh purity argon (Ar) was used as a buffer gas and diethylzinc (DEZn) as a reacting molecular gas sources. The cone angle and apex diameter of the fiber probe used for NFO-CVD were 30° and less than 30 nm, respectively. Since a bare fiber probe without an opaque coating was used for deposition, far-field light leaked to the circumference of the fiber probe and an optical near field (ONF) was generated at the apex. This allowed us to investigate the deposition by an optical far field and ONF simultaneously. The optical power from the fiber probe was measured with a photodiode placed behind

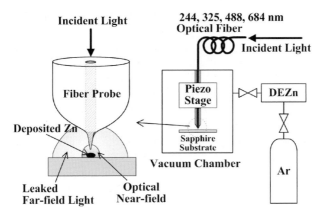

Fig. 23. Experimental setup for CVD using an optical near field

the sapphire substrate. The deposited Zn dots were measured using a shear-force microscope with the fiber probe used for deposition. The buffer gas was ultrahigh purity argon (Ar) at 3 Torr and the gas source of reactant molecules was diethylzinc (DEZn) at 100 mTorr at room temperature. The second harmonic ($\hbar\omega = 5.08\,\text{eV}$) of an Ar$^+$ laser was used as a light source that resonates the absorption band of DEZn. An He–Cd laser (3.81 eV) was used as a nearly resonant light source with the absorption band edge, E_{abs} (4.13 eV) of DEZn [45]. Ar$^+$ (2.54 eV) and diode (1.81 eV) lasers were used as nonresonant light sources.

Figure 24 shows shear-force topographical images of the sapphire substrate after NFO-CVD using the ONF for 5.08 eV (a), 3.81 eV (b), 2.54 eV (c), and 1.81 eV (d). The laser power and irradiation time were (a) 1.6 µW and 60 s, (b) 2.3 µW and 60 s, (b) 360 µW and 180 s, and (c) 1 mW and 180 s, respectively. While the previous work using conventional far-field optical CVD has claimed that a Zn film cannot be grown using nonresonant light ($\hbar\omega < 4.13\,\text{eV}$) [71], we observed the deposition of Zn dots on the substrate just below the apex of the fiber probe using NFO-CVD, even with nonresonant light. The chemical composition of the deposited material was confirmed by X-ray photoelectron spectroscopy. Moreover, we observed luminescence from nanometric ZnO dots, which were prepared by oxidizing the Zn dots fabricated by NFO-CVD [43]. These experimental results imply that the Zn was very pure.

In Fig. 24a and b, the photon energies exceed the dissociation energy ($E_d = 2.26\,\text{eV}$) of DEZn, and exceed E_{abs} for Fig. 24a and is close to the E_{abs} of DEZn for Fig. 24b, i.e., $\hbar\omega > E_d$ and $\hbar\omega \geq E_{abs}$ [45], respectively. The diameter (FWHM) and height of the topographical images were 45 and 20 nm for Fig. 24a and 45 and 26 nm for Fig. 24b, respectively. These images have wide based, as shown by the dotted curves. These bases are a Zn layer, more than 5-nm thick for Fig. 24a and less than 2-nm thick for Fig. 24b, which are deposited by far-filed light leaking from the bare fiber probe. This deposition

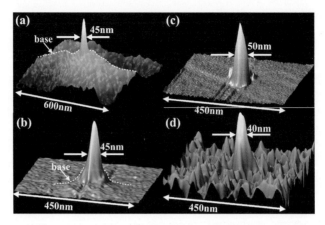

Fig. 24. Shear-force topographical images after NFO-CVD at photon energies of $\hbar\omega = 5.08\,\text{eV}$ (**a**), $3.81\,\text{eV}$ (**b**), $2.54\,\text{eV}$ (**c**), and $1.81\,\text{eV}$ (**d**). The scanning areas are $600 \times 600\,\text{nm}$ for (**a**) and $450 \times 450\,\text{nm}$ for (**b**)–(**d**). The observed laser output power and the irradiation time for deposition were $1.6\,\mu\text{W}$ and $60\,\text{s}$ (**a**), $2.3\,\mu\text{W}$ and $60\,\text{s}$ (**b**), $360\,\mu\text{W}$ and $180\,\text{s}$ (**c**), and $1\,\text{mW}$ and $180\,\text{s}$ (**d**), respectively

is possible because DEZn can absorbs the used lights ($\hbar\omega = 5.08$ and $3.81\,\text{eV}$). The very high peak in the image suggests that ONF enhances the photodissociation rate, because ONF intensity increases rapidly near the apex of the fiber probe.

In Fig. 24c, the photon energy still exceeds the dissociation energy of DEZn, but it is lower than the absorption edge, i.e., $\hbar\omega > E_d$ and $\hbar\omega < E_{\text{abs}}$ [45,72]. The diameter and height of the image were 50 and 24 nm, respectively. While high intensity far-field light leaked from the bare fiber probe, it did not deposit a Zn layer, so there is no foot at the base of the peak. This confirmed that the photodissociation of DEZn and Zn deposition occurred only with an ONF of $\hbar\omega > 2.54\,\text{eV}$.

In Fig. 24d, $\hbar\omega < E_d$ and $\hbar\omega < E_{\text{abs}}$. Even with such low photon energy, we succeeded in depositing of Zn dots. The topographical image had a diameter and height of 40 and 2.5 nm, respectively. We claim that these depositions of Zn dots, in Fig. 24b–c, are peculiar phenomena to an ONF, because while high intensity far-field light leaked from the bare fiber probe, the Zn dots are deposited on the substrate just below the apex of the fiber probe.

To quantify this novel photodissociation process (Figs. 24b–c), we examine the relationship between the photon flux, I, and the deposition rate of Zn, R, in Fig. 25. For $\hbar\omega = 3.81\,\text{eV}$ ($\triangle$), R is proportional to I. For $2.54\,\text{eV}$ ($\Diamond$) and $1.81\,\text{eV}$ ($\circ$), higher-order dependencies appear and are fitted by the third-order function $R = a \cdot I + b \cdot I^2 + c \cdot I^3$. The respective values of $a_{\hbar\omega}$, $b_{\hbar\omega}$, and $c_{\hbar\omega}$ are $a_{3.81} = 5.0 \times 10^{-6}$ and $b_{3.81} = c_{3.81} = 0$ for $\hbar\omega = 3.81\,\text{eV}$; $a_{2.54} = 4.1 \times 10^{-12}$, $b_{2.54} = 2.1 \times 10^{-27}$ and $c_{2.54} = 1.5 \times 10^{-42}$ for $\hbar\omega = 2.54\,\text{eV}$; $a_{1.81} = 0$,

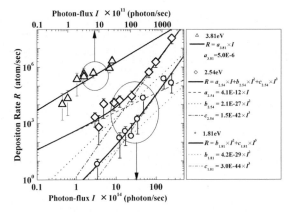

Fig. 25. The optical power (photon-flux: I) dependency of the rate R of Zn deposition. The *solid curves* fit the results using $R = a \cdot I + b \cdot I^2 + c \cdot I^3$

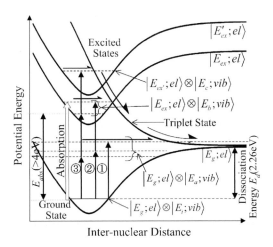

Fig. 26. Schematic drawing of potential curves of an electron in DEZn molecular orbitals. The relevant energy levels of the molecular vibration modes are indicated by the horizontal *broken lines*

$b_{1.81} = 4.2 \times 10^{-29}$, and $c_{1.81} = 3.0 \times 10^{-44}$ for $\hbar\omega = 1.81\,\mathrm{eV}$. These values are used to investigate the physical origin of nonresonant NFO-CVD.

Figure 26 schematically shows the potential curves of an electron in a DEZn molecular orbital drawn as a function of the internuclear distance of the C–Zn bond, which is involved in photodissociation [45]. The relevant energy levels of the molecular vibration mode are indicated by the horizontal broken lines in each potential curve. When a far-field light is used, photoabsorption (indicated by the white arrow in this figure) triggers the dissociation of DEZn [73]. By contrast, when nonresonant ONF is used, there are three possible origins of photodissociation, which we describe as follows [21]. They are (1) the multiple photon absorption process, (2) a multiple step transition process via the intermediate energy level induced by the fiber probe, and (3) the multiple step transition via an excited state of the molecular vibration mode. Possibility (1) is negligible, because the optical power density in the experiment was less than $10\,\mathrm{kW\,cm^{-2}}$, which is too low for multiple photon absorption [20]. Possibility (2) is also negligible, because NFO-CVD was observed for the light in the ultraviolet–near-infra red region, although DEZn lacks relevant energy levels for such a broad region. As a result, our experimental results strongly supported possibility (3), i.e., the physical origin of the photodissociation caused by nonresonant ONF is a transition to an excited state via a molecular vibration mode. The three multiple-step excitation processes in Fig. 26, labeled by ①, ②, and ③, contributed to this.

To evaluate these contributions, we propose the exciton–phonon polariton (EPP) model, which describes the ONF generated at the nanometric probe tip [26]. ONF is a highly mixed state with material excitation rather than a propagating light field; particularly, electronic excitation near the probe tip driven by photons incident into the fiber probe causes mode–mode or anharmonic couplings of phonons. They are considered renormalized phonons, which allow multiple-phonon transfer from the tip to a molecule simultaneously.

6.2 Exciton–Phonon Polariton Model

In the EPP model, the ONF is able to excite the molecular vibration mode due to the steep spatial gradient of the ONF. Figure 27 illustrates the excitation of the molecular vibration mode by ONF and EPP schematically. For an optical far field, the field intensity is uniform in a neutral molecule smaller than the wavelength. Only the electrons in the molecule respond to the electric field with the same phase and intensity, as shown in Fig. 27a. Therefore, an optical far field cannot excite the molecular vibration. By contrast, the field

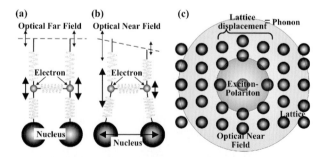

Fig. 27. Schematic explanations of the response of molecule for an optical far field (**a**) and for an optical near field (**b**). (**c**) Schematic description of an exciton–phonon polariton

intensity is not uniform in a molecule for an ONF with a steep spatial gradient. The electrons respond nonuniformly, and the molecular vibration modes are excited because the molecular orbital changes and the molecule is polarized as a result of this nonuniform response of the electrons, as shown in Fig. 27b. We propose the EPP model to quantify this excitation process. The EPP is a quasi-particle, which is an exciton polariton trailing the phonon (lattice vibration) generated by the steep spatial gradient of its optical filed, as shown in Fig. 27c. The EPP model is formulated later.

The model Hamiltonian for the ONF probe can be diagonalized using the conventional theory [74,75], and expressed in such a quasi-particle (EPP) representation as $H = \sum_p \hbar\omega(p)\xi_p^\dagger\xi_p$. Here, the creation (annihilation) operator for EPP and the frequency are denoted $\xi_p^\dagger(\xi_p)$ and $\omega(p)$, respectively. Therefore, in this model, a molecule located near the probe tip does not absorb simple photons, but absorbs EPP, the energies of which are transferred to the molecule, exciting molecular vibrations or induces electronic transitions.

We will now discuss the dissociation probability of a molecule, assuming that the deposition rate of the metal atoms is proportional to the molecular dissociation rate. The transitions from the initial to the final states can be formulated accordingly to the conventional perturbation method for the interaction Hamiltonian that is given by the multipolar QED Hamiltonian in the dipole approximation [76] for an optical near field-molecule interaction as

$$H_{\text{int}} = -\int \boldsymbol{\mu}(\boldsymbol{r}) \cdot \boldsymbol{D}^\perp(\boldsymbol{r})\, \mathrm{d}^3 r \, , \tag{4}$$

$$\boldsymbol{D}^\perp(\boldsymbol{r}) = \mathrm{i}\sum_p \left(\frac{2\pi\hbar\omega_p}{V}\right)^{1/2} \boldsymbol{\varepsilon_p} \left[a_p \exp(\mathrm{i}\boldsymbol{p}\cdot\boldsymbol{r}) - a_p^\dagger \exp(-\mathrm{i}\boldsymbol{p}\cdot\boldsymbol{r})\right] \, . \tag{5}$$

Here $\boldsymbol{\mu}(\boldsymbol{r})$ and $\boldsymbol{D}^\perp(\boldsymbol{r})$ denote the electric-dipole operator and the electric-displacement vector at position r, respectively. The polarization unit vector of a photon is designated as $\boldsymbol{\varepsilon_p}$. Rewriting the photon operators $(a_p, a_p^\dagger)$ in terms of the exciton–phonon polariton operators $(\xi_p, \xi_p^\dagger)$ discussed above, and noticing that the electric-dipole operator consists of two components (electronic and vibrational), we have the interaction Hamiltonian expressed in terms of EPP as

$$H_{\text{int}} = \mathrm{i}\left\{\mu^{\text{el}}\left(e + e^\dagger\right) + \mu^{\text{nucl}}\left(v + v^\dagger\right)\right\}$$
$$\times \sum_p \sqrt{\frac{2\pi\hbar\omega_p}{V}}\left\{v_p v_p'\left(\xi_p + \xi_p^\dagger\right)\right\} e^{\mathrm{i}\boldsymbol{p}\cdot\boldsymbol{r}} \, . \tag{6}$$

Here, μ^{el} and μ^{nucl} are the electronic and vibrational dipole moments, respectively, and the creation (annihilation) operators of the electronic and vibrational excitations are denoted $e^\dagger(e)$ and $v^\dagger(v)$, respectively. The incident photon frequency and transformation coefficients are ω_p and $v_p(v_p')$, respectively. Then, the transition probability of one-, two-, and three-step excitation (labeled ①, ②, and ③ in Fig. 26, and denoted by the corresponding final states as $|f_{\text{first}}\rangle$, $|f_{\text{second}}\rangle$, and $|f_{\text{third}}\rangle$) can be written as

$$
\begin{aligned}
P_{\text{first}}(\omega_{\text{p}}) &= \frac{2\pi}{\hbar}\left|\langle f_{\text{first}}|H_{\text{int}}|\text{i}\rangle\right|^2 \\
&= \frac{(2\pi)^2}{\hbar d}v_{\text{p}}^2 v_{\text{p}}'^2 u_{\text{p}}'^2\left(\mu^{\text{nucl}}\right)^2(\hbar\omega_{\text{p}})I_0(\omega_{\text{p}}),
\end{aligned} \tag{7}
$$

$$
\begin{aligned}
P_{\text{second}}(\omega_{\text{p}}) &= \frac{2\pi}{\hbar}\left|\langle f_{\text{second}}|H_{\text{int}}|\text{i}\rangle\right|^2 \\
&= \frac{(2\pi)^3}{\hbar d^2}\frac{v_{\text{p}}^4 v_{\text{p}}'^6 u_{\text{p}}'^2}{\left|\hbar\omega(p)-(E_{\text{a}}-E_{\text{i}}+\text{i}\gamma_{\text{m}})\right|^2}\left(\mu^{\text{el}}\right)^2\left(\mu^{\text{nucl}}\right)^2(\hbar\omega_{\text{p}})^2 \\
&\quad\times I_0(\omega_{\text{p}})^2,
\end{aligned} \tag{8}
$$

$$
\begin{aligned}
P_{\text{third}}(\omega_{\text{p}}) &= \frac{2\pi}{\hbar}\left|\langle f_{\text{third}}|H_{\text{int}}|\text{i}\rangle\right|^3 \\
&\quad\times\frac{(2\pi)^4}{\hbar d^3}\frac{v_{\text{p}}^6 v_{\text{p}}'^{10} u_{\text{p}}'^2}{\left|\hbar\omega(p)-(E_{\text{a}}-E_{\text{i}}+\text{i}\gamma_{\text{m}})\right|^2\left|\hbar\omega(p)-(E_{\text{ex}}-E_{\text{g}}+\text{i}\gamma_{\text{m}})\right|^2} \\
&\quad\times\left(\mu^{\text{el}}\right)^4\left(\mu^{\text{nucl}}\right)^2(\hbar\omega_{\text{p}})^3 I_0(\omega_{\text{p}})^3,
\end{aligned} \tag{9}
$$

where d, u_{p}, and $I_0(\omega_{\text{p}})$ represent the probe-tip size, the transformation coefficient, and incident light intensity, respectively. Energy conservation is assumed in each transition probability. For this purpose, the following initial and three final states of a system consisting of the optical near-field probe and a molecule are prepared:

$$
\begin{aligned}
|\text{i}\rangle &= |\text{probe}\rangle\otimes|E_{\text{g}};\text{el}\rangle\otimes|E_i;\text{vib}\rangle, \\
|f_{\text{first}}\rangle &= |\text{probe}\rangle\otimes|E_{\text{g}};\text{el}\rangle\otimes|E_a;\text{vib}\rangle, \\
|f_{\text{second}}\rangle &= |\text{probe}\rangle\otimes|E_{\text{ex}};\text{el}\rangle\otimes|E_b;\text{vib}\rangle, \\
|f_{\text{third}}\rangle &= |\text{probe}\rangle\otimes|E_{\text{ex}'};\text{el}\rangle\otimes|E_c;\text{vib}\rangle,
\end{aligned} \tag{10}
$$

where $|\text{probe}\rangle$, $|E_{\alpha};\text{el}\rangle$, and $|E_{\beta};\text{vib}\rangle$ represent a probe state, molecular electronic and vibrational states, respectively. In addition, E_{α} ($\alpha=\text{g,ex,ex}'$) and E_{β} ($\beta=i,a,b,c$) represent the molecular electronic and vibrational energies, respectively, as schematically shown in Fig. 26, and γ_{m} and γ_{m}' are the linewidth of the vibrational and electronic states, respectively. It follows that these near-resonant transition probabilities have the following ratio:

$$
\begin{aligned}
\frac{P_{\text{second}}(\omega_{\text{p}})/I_0(\omega_{\text{p}})^2}{P_{\text{first}}(\omega_{\text{p}})/I_0(\omega_{\text{p}})} &= \frac{P_{\text{third}}(\omega_{\text{p}})/I_0(\omega_{\text{p}})^3}{P_{\text{second}}(\omega_{\text{p}})/I_0(\omega_{\text{p}})^2} \\
&= \frac{\hbar}{2\pi}\frac{P_{\text{first}}(\omega_{\text{p}})}{|\gamma_{\text{m}}|^2 I_0(\omega_{\text{p}})}\left(\frac{v_{\text{p}}'}{u_{\text{p}}'}\right)^2\left(\frac{\mu^{\text{el}}}{\mu^{\text{nucl}}}\right)^2.
\end{aligned} \tag{11}
$$

Here we assumed $\gamma_{\text{m}}=\gamma_{\text{m}}'$, for simplicity. Using this ratio, we analyze the experimental intensity dependence of the deposition rate to clarify possibility (3). For $\hbar\omega=2.54\,\text{eV}$, all the processes (①, ②, and ③) depicted in Fig. 26 are possible, because $\hbar\omega>E_{\text{d}}$ (although $\hbar\omega<E_{\text{abs}}$). Fitting the experimental

value of $P_{\text{first}}(\omega_{2.54}) = a_{2.54}I_0(\omega_{2.54}) = 10^2$ events s^{-1} with reasonable values of $\mu^{\text{nucl}} = 1\,\text{Debye}$, $\mu^{\text{el}} = 10^{-3}\,\text{Debye}$, $\gamma_m = 10^{-1}\,\text{eV}$, $(v'_{\text{p}}/u'_{\text{p}})^2 = 0.01$, and $d = 30\,\text{nm}$ we obtain the following value for the ratio:

$$\frac{P_{\text{second}}(\omega_{2.54})/I_0(\omega_{2.54})^2}{P_{\text{first}}(\omega_{2.54})/I_0(\omega_{2.54})} = \frac{P_{\text{third}}(\omega_{2.54})/I_0(\omega_{2.54})^3}{P_{\text{second}}(\omega_{2.54})/I_0(\omega_{2.54})^2}$$

$$= \frac{\hbar}{2\pi} \frac{P_{\text{first}}(\omega_{2.54})}{|\gamma_m|^2 I_0(\omega_{2.54})} \left(\frac{v'_{\text{p}}}{u'_{\text{p}}}\right)^2 \left(\frac{\mu^{\text{el}}}{\mu^{\text{nucl}}}\right)^2 \simeq 10^{-15},$$

(12)

which is in good agreement with the experimental values in Fig. 25

$$b_{2.54}/a_{2.54} \simeq c_{2.54}/b_{2.54} \simeq 10^{-15}.$$

(13)

For $\hbar\omega = 1.81\,\text{eV}$, dissociation occurs via either ② or ③ shown in Fig. 26, because $\hbar\omega < E_{\text{d}}$ (E_{abs}). The ratio can be evaluated as

$$\frac{P_{\text{third}}(\omega_{1.81})/I_0(\omega_{1.81})^3}{P_{\text{second}}(\omega_{1.81})/I_0(\omega_{1.81})^2}$$

$$= \frac{\hbar}{2\pi} \frac{P_{\text{first}}(\omega_{1.81})}{|\gamma_m|^2 I_0(\omega_{1.81})} \left(\frac{v'_{\text{p}}}{u'_{\text{p}}}\right)^2 \left(\frac{\mu^{\text{el}}}{\mu^{\text{nucl}}}\right)^2 \simeq 10^{-15},$$

(14)

which is also in good agreement with the experimental value $c_{1.81}/b_{1.81} \simeq 10^{-15}$. For the theoretical estimation, we use the experimental value for $P_{\text{first}}(\omega_{1.81}) \simeq a_{2.54}I_0(\omega_{2.54}) = 10^2$ events s^{-1} because both transitions for light with photon energies of 1.84 and 2.54 eV are attributed to the coupling between phonons in the probe and molecular vibrations. The overall agreement between the theoretical and experimental results suggests that the EPP model provides a way to understand the physical origin of the near-field photodissociation process. For $\hbar\omega = 3.81\,\text{eV}$, the direct absorption by the electronic state is much stronger than in other cases, because the light is near resonant for DEZn. This is why we did not observe higher-order power dependence of the deposition rate in the optical power region that we observed.

To clarify the physical origin of this process, the optical power and photon energy dependencies of the deposition rates were measured, and we explain the dependencies using a multiple-step excitation process via the molecular vibration mode and the EPP model. In this model, the enhanced coupling between the optical field and molecular vibration originates from the steep spatial gradient of the optical power of ONF. Such a nonadiabatic photochemical process violates the Franck–Condon principle that assumes an adiabatic process, and is applicable to other photochemical phenomena.

6.3 Nonadiabatic Near-Field Photolithography

According to the EPP model, the nonadiabatic photochemical process is a universal phenomenon applicable to many other photochemical processes, including photoresist exposure. Following from this model, we proposed a novel method of photolithography that uses a nonadiabatic photochemical process, i.e., nonadiabatic near-field photolithography. Using this method, UV–photoresist, which is suitable for nanolithography, can be exposed using inexpensive visible light sources and equipment without using expensive UV light sources.

The wave properties of light can cause problems for nanometric photolithography, including not only the diffraction limit, but also coherency and polarization dependence. In photolithography of high-density nanometric arrays, the optical coherent length is longer than the separation between adjacent corrugations, even when an Hg lamp is used, and there is not enough photoresist absorption to suppress fringe interference of scattered light due to the narrow separation. The transmission intensity of light passing through a photomask strongly depends on its polarization, so the design of photomask structures must include such effects. Since the optical near field in nonadiabatic near-field photolithography has no wave properties, these problems are easily solved. In this study, we attempted nanophotolithography using visible 672- and 532-nm lasers for the UV photoresist; this paper illustrates the exposure time dependence of exposure depth. We were able to fabricate T-, L-, and ring-shaped arrays using nonadiabatic near-field photolithography. We were also able to demonstrate, for the first time, nonadiabatic near-field photolithography using a photoinsensitive electron-beam resist.

Figure 28a shows a schematic configuration of the photomask used and the Si-substrate on which the photoresist (OFPR-800 or TDMR-AR87: Tokyo-Ohka Kogyo Co.) was spin-coated. These were used in contact mode, and we kept the gap between the photomask and photoresist as narrow as possible. Figure 28b–d shows AFM images of the photoresist surface after exposure and development. Figure 28b shows that the nonadiabatic photochemical process created a corrugated pattern on the photoresist although photoresist OFPR-800 does not react to propagating 672-nm light. The corrugated pattern was 30 nm deep and 150 nm wide, much smaller than the wavelength of the light source.

To decrease exposure time, we used other light sources and photoresists, keeping in mind the material properties of the nonadiabatic photochemical process and its strong dependence on photon energy [48, 49]. Figure 28c illustrates our use of photoresist TDMR-AR87 with the g-line of an Hg lamp as the light source. The corrugated pattern on the photoresist caused by the nonadiabatic photochemical process was obtained with a 60-s exposure, although TDMR-AR87 does not react to the g-line. Figure 28d shows that a similar pattern was obtained with exposure times as short as 3 s.

Grooves on the photoresist appeared along the edges of the Cr mask pattern. In this region, a steeply spatial gradient of optical energy indicates the existence of an optical near field. Therefore, these results of exposure indicate that the nonadiabatic process originates from an optical near-field effect [48, 49].

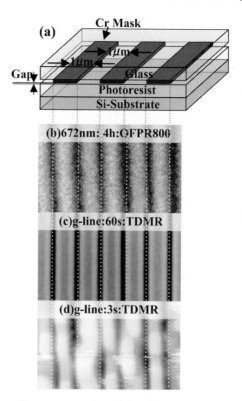

Fig. 28. (**a**) A schematic representation of the photomask and the Si-substrate spin-coated with photoresist (OFPR-800 or TDMR-AR87) during the exposure process. (**b**) Atomic force microscopy images of photoresist OFPR-800 developed after a 4-h exposure to a 672-nm laser. (**c**) Atomic force microscopy images of photoresist TDMR-AR87 exposed to the g-line of an Hg lamp for 30 s; and (**d**) Atomic force microscopy images of photoresist TDMR-AR87 exposed to the g-line of an Hg lamp for 3 s

Figure 29 shows the exposure time dependency of groove depth in the corrugated pattern for two sets of light source and photoresist. The optical power densities of the exposure lights were $30\,\mathrm{mW/cm^2}$ for the g-line light and $1\,\mathrm{W\,cm^{-2}}$ for the 672-nm laser; both dependencies saturated at a depth of about 100 nm. These results support the hypothesis that the nonadiabatic process originates from an optical near-field effect, since the optical near field is localized at the edge of the Cr mask. The exposure rate in the first set of TDMR-AR87 and a g-line light source was more than 10^4 times higher than for the OFPR-800 with a 672-nm laser. This increase in exposure rate drastically decreased the exposure time, making it sufficiently short for mass production.

In demonstrations of conventional nanophotolithography reported by other researchers, the photomask with the lines and spaces structure are popularly

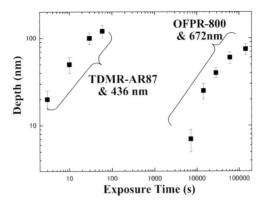

Fig. 29. Depth of the developed grooves versus exposure time

used, because the problems come from the interference fringes and the negative effect of light polarization can be reduced by selecting the polarization. For more general patterns such as L-, T-, and ring-shaped arrays, however, it is impossible to reduce the problem, even if polarization is controlled. Figure 30 shows AFM photoresist images after development. We used photomasks with L-, T-, and solid-circle Cr mask-arrays (see Fig. 30a–c, respectively). At exposure, we used linear polarized light with the Hg g-line ($\lambda = 435\,\mathrm{nm}$) and TDMR-AR87, which is i-line ($\lambda = 365\,\mathrm{nm}$) photoresist. We were able to reproduce the formation of arrays with the expected shapes. For comparison, we exposed the photoresist using the i-line of an Hg lamp, which is used for conventional photolithography. We did not obtain the expected shape, but only a pattern exposed by fringe light interference. Since the TDMR-AR87 photoresist has a low absorbance, there was a strong interference effect. This successful development of arrays with complex structures means that nonadiabatic near-field photolithography can have practical uses.

Since this novel process results from the excitation of molecular vibration modes, the nonadiabatic photochemical process using an optical near field can induce a photochemical reaction even in photoinsensitive organic molecules; electron beam resist is one example.

Figure 31 shows AFM images of the electron beam resist (ZEP520: ZEON) surfaces after exposure and development using nonadiabatic near-field photolithography. The exposure light source was the third harmonic generation of a Q-switched Nd:YAG laser with a repetition rate of 20 Hz and a typical pulse width of 10 ns. The exposure light power density was $20\,\mathrm{mW\,cm^{-2}}$ and the exposure duration was 5 min. The resist thickness was less than 80 nm, and the electron beam resist was prebaked at 180°C for 2 min. We used a photomask with a 2D array pattern of 1 μm diameter Cr-disks (the disks were 100 nm thick and separated by 2 μm). The incident angle of the exposure light was 70°, as shown in Fig. 32.

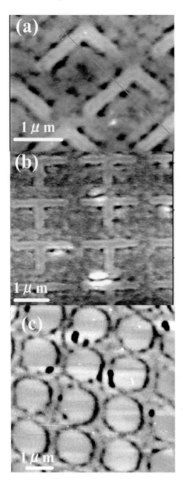

Fig. 30. Atomic force microscopy images after development of photoresist TDMR-AR87 exposed to the g-line: (**a**) L-shaped array, (**b**) T-shaped array, and (**c**) *solid circle*-shaped array

In Fig. 31a, the arrow indicates the direction of the incident light. After development, we obtained fabricated 2D arrays on the photoresist. The nonadiabatic photochemical process exposed even the electron-beam resist. The photoresist was not exposed by direct photoirradiation, and the fabricated structure, which had a diameter of $1\,\mu m$, was asymmetrical because the light incident angle was $70°$ (see Fig. 32). Therefore, this result strongly supports the hypothesis that exposure of electron beam resists is due to the nonadiabatic photochemical process; the optical near-field distribution was not cylindrically symmetrical due to the large incident angle of light, as shown in Fig. 32b. Figure 31b shows cross-sectional profiles of the patterns developed across broken lines A and B in Fig. 31a. The pit width was $50\,nm$, which is

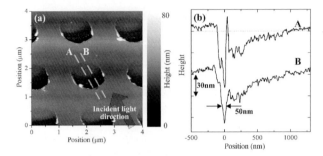

Fig. 31. (a) Atomic force microscopy images after developing electron beam resist exposed using a 355-nm laser. (b) Cross-sectional profiles of the developed pattern along *broken lines* A and B in Fig. 27a

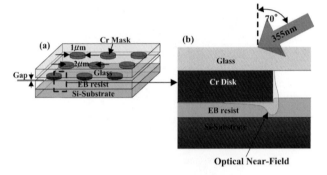

Fig. 32. (a) Schematic configuration of the photomask used and the Si-substrate spin-coated with electron beam resist (ZEP520) during the exposure process. (b) Magnified schematic drawing of the contact region. The optical near field is eliminated, as shown in the drawing, due to the low incident angle of the exposure light

much smaller than the wavelength of the incident lights. The pit depth was 70 nm, deep enough to reach the Si substrate. The structure fabricated on the electron beam resist had a keen edge, providing a higher fabrication resolution, compared with fabricated structures using a photoresist (see Figs. 28 and 30). The great advantages of this keen edge are due to the properties of the electron beam resist, including a greater uniformity and smoother surface.

Using nonadiabatic near-field photolithography, we succeeded in exposing UV photoresist using a visible light source, while the photoresist is inactive to visible light. Optical near-field features enable fabrication of nanometric 2D arrays with complex shapes. Experimental results support the hypothesis that nonadiabatic photolithography is suitable for developing actual electronic and photonic devices. Finally, we succeeded in exposing the electron beam resist using nonadiabatic photolithography. Since the properties of the electron beam resist make it suitable for developing fine structures, we obtained nanometric structures: keen-edged nanopits 50 nm wide and 70 nm deep.

7 A Control of an Optical Near Field Using a Fiber Probe

7.1 Second Harmonic Generation in an Al-Coated Probe

First, we review the experimental results about the second harmonic generation (SHG), i.e., a wavelength conversion of an optical near filed, in a fiber probe coated with Al. Figure 33a shows the cross-sectional profile of the employed fiber probe for SHG (probe A) and schematic explanation of the probe-to-probe experiment for measuring the spatial distribution of the second harmonic (SH) intensity by using the probe S. A triple-tapered fiber probe, employed for the ultraviolet region [77], was used as the probe A. The fiber had the double core structure consisting of a GeO_2 doped core and a pure silica core, whose diameters were 160 nm and 3.0 μm, respectively. In the fabrication of the probe A, we used two kinds of a buffered hydrogen fluoride (BHF) solution as etchant. The two solutions differ in the volume ratios of NH_4F solution (40 wt%):HF acid (50 wt%):H_2O, which were 1.7:1:1 (BHF_A) and 10:1:1 (BHF_B), respectively. First, the fiber end was etched by BHF_A for 56 min to taper the fiber. Next, the tapered fiber was etched by BHF_B for 30 min to sharpen the center of the fiber core. The cone angle of the fabricated fiber probe was 17°. The triple-tapered end was coated with 500 nm thick aluminum by vacuum evaporation. Metal coating was performed in two different ways: either coating the aluminum from the top of the probe (probe A) or from the lateral side of the probe (Probe B). Lower part in Fig. 33a shows the SEM image of the top view of an aluminum coated sharpened fiber for the probe A. The aperture of the probe A was created by using the focused ion beam technique. Aperture diameter of 100 nm for probe A is clearly evident from the SEM image shown in Fig. 33a.

Figure 33b shows a schematic drawing of the experimental setup for measuring SH intensity. The SH intensity was measured using a cleaved fiber.

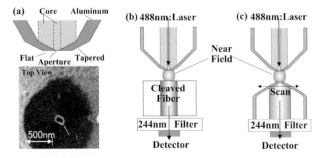

Fig. 33. Schematic explanation of experiments and fiber probe. (**a**) A cross-sectional profile and an SEM image of the near-field probe for SHG. (**b**) The schematic drawing of experimental configuration for measurement of SH intensity from the probe. (**c**) The schematic drawing of the probe-to-probe experiment

The 10-μm core of the cleaved fiber was much larger than the apex diameter of the probe and the distance between the cleaved fiber and the probe was less than 1 μm. Therefore, most of the fundamental and SH signals from the probe could be detected by the cleaved fiber. Figure 33c explains the method used for measuring the spatial distribution of the SH intensity in the probe-to-probe experiment [78]. The second probe used in the probe-to-probe scanning was a single tapered fiber probe that was fabricated by pulling and etching a fiber with a pure silica core that was then coated with a 500-nm thick layer of aluminum [2]. The diameter of its aperture was 100 nm. Separation between the two probes was regulated to several nanometers by using a shear-force technique. A 488-nm wavelength Ar$^+$ laser was used as the light source for the experiments. An R166UH (Hamamatsu Photonics Co. LTD) photomultiplier tube was used to detect the SH signal selectively because it is sensitive in the wavelength range from 160 to 320 nm. In addition, a band-pass filter with a bandwidth of 10 nm was used to reject the fundamental signal with an extinction ratio higher than 10^{-4} for the fundamental wavelength. The quantum efficiency of the photomultiplier tube was more than 30% for 244 nm light. The photon counting technique was used for detection. For an accumulation time of 1,000 s, the lowest observable SH power was 3.0×10^{-19} W.

Figure 34 shows the relationship between the fundamental and SH powers when we used the fundamental signal passing through the probe as the reference for this measurement. The squares correspond to the experimental results for Probe A. The SHG power is proportional to the square of the fundamental power in the low power region and saturates at about a fundamental power of 20 nW. Since the fiber probe can be thermally damaged by a fundamental output exceeding 100 nW, the observed saturation at 20 nW is reasonable. The thermal damage changes condition of the fiber probe, such as an aperture diameter, an aluminum coating, and so on, then the SH signal intensity is saturated. The solid line fits the measured values in the unsaturated region.

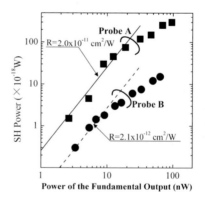

Fig. 34. Relation between the fundamental and SH powers. *Squares* and *circles* represent the experimental results of the probes A and B, respectively. The solid and broken lines are fitted to them in the unsaturated region

The conversion factor R is the ratio of the SH intensity to the square of the fundamental intensity. From the slope of the solid line, the value of R for Probe A with an aperture of 100 nm is 2.0×10^{-11} cm^2 W^{-1}. Since this value depends on the aluminum deposition technique, SHG was also evaluated for Probe B, which was fabricated by the alternate approach of depositing aluminum from the side of the triple-tapered fiber. The results are shown by circles in Fig. 34, and the broken line fits the data points for Probe B in the unsaturated region. The slope of this line gives the R value for Probe B, which is estimated to be 2.1×10^{-12} cm^2 W^{-1}, ten times smaller than the value for Probe A. Such a large difference in the value of R supports the dependence of SHG on the deposition technique of coatings. Furthermore, it should be noted that the values of R for Probes A and B are five orders higher than that obtained for aluminum surface coated on large flat pieces of glass. Such a large value corresponds to the value for a 5-mm thick KDP crystal in the far-field configuration [20].

Figure 35a shows the spatial distribution of the SH power on Probe A obtained in the probe-to-probe experiment. The image size is $1.5 \times 1.5\,\mu$m. The solid ellipse (X) shows the position of the aperture, while the broken curve (Y) shows the boundary between the flat and the tapered areas. Strong SH light is emitted at the areas enclosed by the dotted dumbbell curve and around the boundary (Y). Figure 35b shows the cross-sectional profiles of SH power along the solid (A) and broken (B) lines in Fig. 35a. The inset shows the top-view SEM image of the probe and the image area corresponds to that of Fig. 35a. The peak power of SH signal was 0.4 pW. This observed power is much larger than the previous experimental result, shown in Fig. 34. We consider this enhancement of SH signal comes from the resonant effect of two probes. In the area enclosed by the dotted dumbbell curve, a strong SH signal was recorded from both sides of the aperture. The SHG observations within the area enclosed by the dumbbell curve may be related to polarization of the SH light and correspond to the propagation mode of the fiber. Near the boundary between the flat and tapered areas (Y), nonlinear oscillation of the surface plasmon in the aluminum coating could easily emit SH light due to the edge effect. These results confirm that the SH signal was generated in the aluminum coating. Investigations of SHG in a bare fiber probe without aluminum coating revealed no detectable SH signals despite a high fundamental intensity. This also strongly indicates that the aluminum coating enhanced SHG in the fiber probe. In a subwavelength-size fiber probe, the optical near field becomes dominant and can excite the surface plasmon polariton on the coated metal, an effect that is well known in the case of flat metal surfaces [79]. Therefore, we believe that the SHG enhancement with the fiber probe occurs as a result of the optical nonlinear response of the surface plasmon polaritons on the coated metal. Note that such SHG effect is negligible small for the nonadiabatic nanofabrication in Sect. 6, since we never observed the SH light from the bare fiber probe using for the nonadiabatic nanofabrication, as we mentioned above.

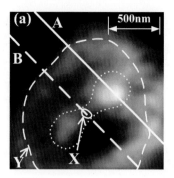

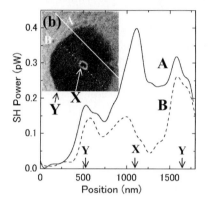

Fig. 35. Spatial distribution of the SH power on the Probe A measured by the probe-to-probe experiment. (**a**) 2D profile of SH distribution. A *solid ellipse* (X) represents the position of the aperture. A *broken curve* (Y) refers to the boundary between the flat and tapered area as is shown in Fig. 33a. (**b**) Cross-sectional profiles of the distribution along the *solid* (A) and *broken* (B) *lines* marked in (**a**). *Inset* shows the SEM image of the probe corresponding to (**a**)

7.2 Gigantic Optical Mangeto-Effect in an Optical Near-Field Probe Coated with Fe

Here, we demonstrate a polarization control of an optical near field using a fiber probe coated with Fe. Figure 36 shows typical SEM images of fabricated optical fibers that were sharpened by etching and coated with Fe and a multilayer metal coating. First, we fabricated a fiber tip coated only with a 150-nm thick Fe layer. When the aperture diameter is much smaller than the diffraction limit of light, the majority of the incident optical power is transferred by a plasmon wave at the boundary between the glass fiber and the Fe layer. The plasmon wave is reconverted to an optical field at the aperture. Therefore, the polarization of the output light passing through the fiber tip can be controlled by the magneto-optical effect of Fe. The detailed fabrication procedure was reported in [58]. We also fabricated a probe tip coated

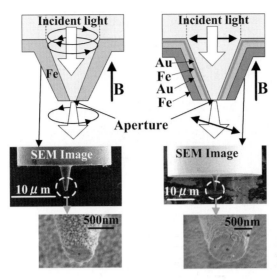

Fig. 36. Schematic structures of Fe-coated fiber probes and SEM images of typical fabricated probes

with Au/Fe/Au/Fe layers (10, 1, 80, and 100, respectively), with the intention of increasing the magneto-optical effect by using the quantum size effect of Au/Fe/Au QW structures [80]. The 100-nm Fe layer, which is outside the QW structure, gives the magnetic coercivity for the probe. The apertures of the probes were fabricated by using a focused ion beam (FIB).

In the experiment, we used an He–Ne laser ($\lambda = 632.8\,\mathrm{nm}$) as the light source; the total fiber length was 5 cm. Fe has the greatest magneto-optical effect on light in this wavelength region [56]. To obtain the MCD and Faraday rotation angle of light output from the probe, the polarization of incident light was adjusted without an external magnetic field by the wave plates and a polarizer set in front of the fiber coupler. An external magnetic field, B, was applied with a permanent magnet; the direction of the field was parallel to the fiber axis. The circular polarization of the output light was measured using a set of quarter-wave plates and a polarizer for MCD measurements and the angle of the polarizer for Faraday rotation measurements, in front of the photodetector.

Figure 37a shows the experimental results of MCD measurements. Here, we used probes with a 50-nm diameter aperture. The degree of polarization of the output light from the probe coated with Fe increased with an increase in the intensity of the external magnetic field, shown by closed circles and squares. However, the degree of polarization of the conventional near-field probe coated with 160-nm thick Au was independent of magnetic field intensity, and was close to 0, shown by closed triangles. This indicates that the MCD of an Fe-coated probe comes from the MCD of Fe. The MCD of the probe coated

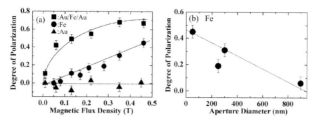

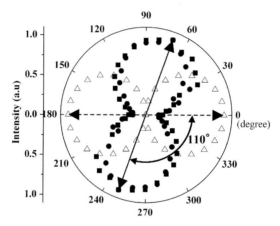

Fig. 37. (a) The magnetic flux density dependence of the degree of polarization in the fabricated probes. (b) The aperture diameter dependence of the degree of polarization with a magnetic field of 0.46 T for an Fe-coated probe

Fig. 38. Polarization of output light from Au/Fe/Au/Fe coated fiber tip with a 50-nm aperture

with Au/Fe/Au/Fe layers was greater than that of the probe coated with only Fe, because of the quantum size effect of Au/Fe/Au QW structures. The difference was especially great in weak magnetic fields. At 0.05 T, the probe coated with Au/Fe/Au/Fe layers had an MCD 40 times greater than that of the probe coated only with Fe. The maximum degree of polarization (0.68) was observed at a magnetic flux density of 0.35 T.

Figure 37b shows that an Fe-coated probe's aperture diameter depends on the degree of polarization with a magnetic field of 0.46 T. The degree of polarization decreases as the aperture diameter increases, because output light from the probe acts as propagation light in the probe tip with an large aperture, and propagation light has a relatively small interaction with the Fe coating layer.

Figure 38 shows the experimental results of the Faraday rotation measurement. The polar angle and radial axis show the angle of the polarizer and detected intensity, respectively. The open triangles show the polarization of output light from the Au/Fe/Au/Fe coated probe before the magnetic field was applied. For a probe with an aperture larger than 200 nm, the Faraday rotation angle was small. However, we observed gigantic Faraday rotation angles in a probe with an aperture of less than a 200 nm; this is much smaller than the diffraction limit. We obtained a Faraday rotation angle of 110° for the Au/Fe/Au/Fe-coated probe with a 50-nm diameter aperture at $B = 0.35$ T, shown by closed circles in Fig. 38. This rotation angle is sufficient for an optical isolator and circulator. For a probe with a 300-nm diameter aperture, the Faraday rotation angle was less than 10°. Therefore, the gigantic Faraday rotation occurs in a small region near the apex of the probe, which is subwavelength in size. Since the Au/Fe/Au/Fe-coated probe had a coercively magnetized Fe layer, it did not require an external magnetic field after magnetization. The Faraday rotation angle was preserved even after the external magnetic field was removed, due to its Fe magnetized layer, shown by closed squares in Fig. 38. This is an advantage for integration, because an external magnetic field, which has a significant and negative effect on its surroundings, is not necessary.

These strong magneto-optical effects occur in the subwavelength region near the apex of the probe. Therefore, this technique can be applied and integrated into a nanophotonic device, to act as an optical isolator, circulator, and so on in nanophotonic integrated circuits [1]. It is probable that these magneto-optical effects derive from the large magnitude of the magneto-optical coefficient of Fe; these effects should be applicable to other nanophotonic devices.

Acknowledgments

I thank T. Yatsui of Japan Science and Technology Agency, T.-W. Kim of Kanagawa Academy of Science and Technology, H. Hori, and I. Banno of Yamanashi University for fruitful discussions. This work was carried out at the projects of ERATO and SORST, Japan Science and Technology Agency, since 1998.

References

1. M. Ohtsu, K. Kobayashi, T. Kawazoe, S. Sangu, T. Yatsui: IEEE J. Select. Top. Quant. Electron. **8**, 839 (2002)
2. *Near-Field Nano/Atom Optics and Technology*, ed. by M. Ohtsu. (Springer, Berlin Heidelberg New York Tokyo 1998)
3. T. Saiki, K. Nishi, M. Ohtsu: Jpn. J. Appl. Phys. **37**, 1638 (1998)
4. J.C. Kim, H. Rho, L.M. Smith, H.E. Jackson, S. Lee, M. Dobrowolska, J.M. Merz, J.K. Furdyna: Appl. Phys. Lett. **73**, 3399 (1998)
5. E. Deckel, D. Gershoni, E. Ehrenfreund, D. Spector, J.M. Garcia, P.M. Petroff: Phys. Rev. Lett. **80**, 4991 (1998)
6. M. Nirmal, B.O. Dabbousi, M.G. Bawendi, J.J. Macklin, J.K. Trautman, T.D. Harris, L.E. Brus: Nature **383**, 802 (1996)
7. N.H. Bonadeo, J. Erland, D. Gammon, D. Park, D.S. Katzer, D.G. Steel: Sicence **282**, 1473 (1998)

8. D. Goldhaber-Gordon, H. Shtrikman, D. Mahalu, D. Abusch-Magder, U. Meirav, M.A. Kastner: Nature **391**, 156 (1998)
9. F. Simmel, R.H. Blick, J.P. Kotthaus, W. Wegscheider, M. Bichler: Phys. Rev. Lett. **83**, 804 (1999)
10. L.W. Molenkamp, K. Flensberg, M. Kemerink: Phys. Rev. Lett. **75**, 4282 (1995)
11. M. Taut: Phys. Rev. B **62**, 8126 (2000)
12. F.R. Waugh, M.J. Berry, D.J. Mar, R.M. Westervelt, K.L. Campman, A.C. Gossard: Phys. Rev. Lett. **75**, 705 (1995)
13. G. Burkard, G. Seelig, D. Loss: Phys. Rev. B **62**, 2581 (2000)
14. K. Mukai, S. Abe, H. Sumi: J. Phys. Chem. B **103**, 6096 (1999)
15. C.R. Kagan, C.B. Murray, M. Nirmal, M.G. Bawendi: Phys. Rev. Lett. **76**, 1517 (1996)
16. T. Kawazoe, K. Kobayashi, J. Lim, Y. Narita, M. Ohtsu: Phys. Rev. Lett. **88**, 067404 (2002)
17. S. Sangu, K. Kobayashi, A. Shojiguchi, T. Kawazoe, M. Ohtsu: J. Appl. Phys. **93**, 2937 (2003)
18. K. Kobayashi, S. Sangu, T. Kawazoe, M. Ohtsu: J. Lumin. **112**, 117 (2005)
19. S.A. Crooker, J.A. Hollingsworth, S. Tretiak, V.I. Klimov: Phys. Rev. Lett. **89**, 186802 (2002)
20. A. Yariv: *Introduction to Optical Electronics*, (Holt, Rinehart and Winston, New York 1971)
21. T. Kawazoe, K. Kobayashi, S. Sangu, M. Ohtsu: Appl. Phys. Lett. **82**, 2957 (2003)
22. T. Kawazoe, K. Kobayashi, S. Sangu, M. Ohtsu: J. Microsc. **209**, 261 (2003)
23. M.N. Islam, X.K. Zhao, A.A. Said, S.S. Mickel, C.F. Vail: Appl. Phys. Lett. **71**, 2886 (1997)
24. G. McDermott, S.M. Prince, A.A. Freer, A.M. Hawthornthwaite-Lawless, M.Z. Papiz, R.J. Cogdell, N.W. Isaacs: Nature **374**, 517 (1995)
25. P. Jordan, P. Fromme, H.T. Witt, O. Klukas, W. Saenger, N. Krauss: Nature **411**, 909 (2001)
26. K. Kobayashi, S. Sangu, H. Ito, M. Ohtsu: Phys. Rev. A **63**, 013806 (2001)
27. H. Morkoç: *Nitride Semiconductors and Devices*. (Springer, Berlin Heidelberg New York 1999)
28. S. Nakamura, G. Fasol: *The Blue Laser Diode*. (Springer, Berlin Heidelberg New York 1997)
29. D. Huang, Y. Fu, H. Morkoç: In: *Optoelectronic Properties of Semiconductor Nanostructures*, ed. by T. Steiner. (Artech House, Boston 2004)
30. A.D. Yoffe: Adv. Phys. **42**, 173 (1993)
31. K. Shimada, T. Sota, K. Suzuki: J. Appl. Phys. **84**, 4951 (1998)
32. F. Widmann, B. Daudin, G. Feuillet, Y. Samson, J.L. Rouviere, N. Pelekanos: J. Appl. Phys. **83**, 7618 (1998)
33. Y. Arakawa, T. Someya, K. Tachibana: Phys. Stat. Sol. B **224**, 1 (2001)
34. F. Widmann, J. Simon, N.T. Pelekanos, B. Daudin, G. Feuillet, J.L. Rouviere, G. Fishman: Microelectron. J. **30**, 353 (1999)
35. P. Ramvall, P. Riblet, S. Nomura, Y. Aoyagi, S. Tanaka: J. Appl. Phys. **87**, 3883 (2000)
36. H. Morkoç, A. Neogi, M. Kuball: Mater. Res. Soc. Symp. Proc. **749**, 56743 (2004)
37. A. Neogi, H. Everitt, H. Morkoç: IEEE Trans. Nanotechnol. **2**, 10 (2003)

38. A. Neogi, H. Everitt, H. Morkoç, T. Kuroda, A. Takeuchi: IEEE Trans. Nanotechnol. **4**, 297 (2005)

39. G. Salviati, F. Rossi, N. Armani, V. Grillo, O. Martinez, A. Vinattieri, B. Damilano, A. Matsuse, N. Grandjean: J. Phys.: Condens. Matter **16**, S115 (2004)

40. P.G. Gucciardi, A. Vinattieri, M. Colocci, B. Damilano, N. Grandjean, F. Semond, J. Massies: Phys. Status Solidi B **224**, 53 (2001)

41. V.V. Polonski, Y. Yamamoto, M. Kourogi, H. Fukuda, M. Ohtsu: J. Microsc. **194**, 545 (1999)

42. Y. Yamamoto, M. Kourogi, M. Ohtsu, V. Polonski, G.H. Lee: Appl. Phys. Lett. **76**, 2173 (2000)

43. T. Yatsui, M. Ueda, Y. Yamamoto, T. Kawazoe, M. Kourogi, M. Ohtsu: Appl. Phys. Lett. **81**, 3651 (2002)

44. J.G. Calvert, J.N. Patts Jr.: In: *Photochemistry.* (Wiley, New York 1966)

45. R.L. Jackson: J. Chem. Phys. **96**, 5938 (1992)

46. T. Kawazoe, Y. Yamamoto, M. Ohtsu: Appl. Phys. Lett. **79**, 1184 (2001)

47. M.M. Alkaisi, R.J. Blaikie, S.J. McNab, R. Cheung, D.R.S. Cumming: Appl. Phys. Lett. **75**, 3560 (1999)

48. T. Kawazoe, K. Kobayashi, S. Takubo, M. Ohtsu: J. Chem. Phys. **122**, 024715 (2005)

49. T. Kawazoe, M. Ohtsu: Proc. SPIE **5339**, 619 (2004)

50. T. Matsumoto, M. Ohtsu, K. Matsuda, T. Saiki, H. Saito, K. Nishi: Appl. Phys. Lett. **75**, 3246 (1999)

51. I.I. Smolyaninov, H.Y. Liang, C.H. Lee, C.C. Davis, S. Aggarwal, R. Ramesh: Opt. Lett. **25**, 835 (2000)

52. S. Noda, M. Yokoyama, M. Imada, A. Chutinan, M. Mochizuki: Science **293**, 1123 (2001)

53. T. Saiki, Y. Narita: Jap. Soc. Appl. Phys. Int. **5**, 22 (2002)

54. B.K. Canfield, S. Kujala1, K. Jefimovs, T. Vallius, J. Turunen M. Kauranen: J. Opt. A **7**, S110 (2005)

55. *Landolt–Boenstein: Zahlenwerte und Funktionen II-9, Magnetischen Eigenschaften 1.* (Springer, Berlin Heidelberg New York 1982)

56. G.S. Krinchik, V.A. Artemev: Sov. Phys. JETP **26**, 1080 (1968)

57. P.B. Johnson, R.W. Christy: Phys. Rev. B **9**, 5056 (1974)

58. J. Lim, T. Kawazoe, T. Yatsui, M. Ohtsu: IEICE Trans. Electron. **E85**, 2077 (2002)

59. Z.K. Tang, A. Yanase, T. Yasui, Y. Segawa, K. Cho: Phys. Rev. Lett. **71**, 1431 (1993)

60. N. Sakakura, Y. Masumoto: Phys. Rev. B **56**, 4051 (1997)

61. A.I. Ekimov, Al.L. Eflos, A.A. Onushchenko: Solid State Commun. **56**, 921 (1985)

62. T. Itoh, S. Yano, N. Katagiri, Y. Iwabuchi, C. Gourdon, A.I. Ekimov: J. Lumin. **60&61**, 396 (1994)

63. T. Suzuki, T. Mitsuyu, K. Nishi, H. Ohyama, T. Tomimasu, S. Noda, T. Asano, A. Sasaki: Appl. Phys. Lett. **69**, 4136 (1996)

64. T. Saiki, S. Mononobe, M. Ohtsu, N. Saito, J. Kusano: Appl. Phys. Lett. **68**, 2612 (1996)

65. T. Saiki, K. Matsuda: Appl. Phys. Lett. **74**, 2773 (1999)

66. *Landolt-Bornstein, Physics of II–VI and I–VII Compounds, Semimagnetic Semiconductors 17b.* (Springer, Berlin Heidelberg New York 1982)

67. Y. Masumoto, T. Kawazoe, N. Matsuura: J. Lumin. **76&77**, 189 (1998)
68. A. Neogi, C.W. Lee, H.O. Everitt, T. Kuroda, A. Tackeuchi, E. Yablonovitch: Phys. Rev. B **66**, 153305 (2002)
69. H. Raether: *Surface Plasmons.* Springer Tracts in Modern Physics, Vol. III (Springer, Berlin Heidelberg New York 1988)
70. M. Born, E. Wolf: *Principles of Optics*, 6th ed. (Pergamon Press, Oxford 1983)
71. M. Shimizu, H. Kamei, M. Tanizawa, T. Shiosaki, A. Kawabata: J. Cryst. Growth **89**, 365 (1988)
72. R.L. Jackson: Chem. Phys. Lett. **163**, 315 (1989)
73. H. Okabe: *Photochemistry.* (John Wiley and Sons, New York 1978)
74. J.J. Hopfield: Phys. Rev. **112**, 1555 (1958)
75. A.L. Ivanov, H. Haug, L.V. Keldysh: Phys. Rep. **296**, 237 (1998)
76. D.P. Craig, T. Thirunamachandran: *Molecular Quantum Electrodynamics.* (Dover Publications, New York 1998)
77. S. Mononobe, T. Saiki, T. Suzuki, S. Koshihara, M. Ohtsu: Opt. Commun. **146**, 45 (1998)
78. T. Yatsui, M. Kourogi, M. Ohtsu: Appl. Phys. Lett. **71**, 1756 (1997)
79. V.M. Agranovich, D.L. Mills: *Surface Polaritons, Electromagnetic Wave at Surface and Interface.* (North-Holland, Amsterdam 1982)
80. Y. Suzuki, T. Katayama, S. Yoshida, K. Tanaka, K. Sato: Phys. Rev. Lett. **68**, 3355 (1992)

Architectural Approach to Nanophotonics
for Information and Communication Systems

M. Naruse, T. Kawazoe, T. Yatsui, S. Sangu, K. Kobayashi, and M. Ohtsu

1 Introduction

To accommodate the continuously growing amount of digital data handled in information and communications systems [1], optics is expected to play a wider role in enhancing the overall system performance by performing certain functional behavior [2] in addition to merely serving as the communication medium. In this regard, for example, so-called all-optical packet switching has been thoroughly investigated. Also, the application of inherent optical features, such as parallelism, in computing systems has been investigated [3,4]. However, many technological difficulties remain to be overcome in order to adopt optical technologies in critical information and communication systems: one problem is the poor integrability of optical hardware due to the diffraction limit of light, which is much larger than the gate width in VLSI circuits, resulting in relatively bulky hardware configurations.

Nanophotonics, on the other hand, which is based on local electromagnetic interactions between nanometer-scale elements, such as quantum dots, via optical near fields, provides ultra high-density integration since it is not constrained by the diffraction limit of light [5]. From an architectural perspective, this drastically changes the fundamental design rules of functional optical systems. Consequently, suitable architectures may be built to exploit this capability of the physical layer. In this chapter, we approach nanophotonics from a system architecture perspective, considering the unique physical principles provided by optical near-field interactions and the functionality required for practical applications.

This chapter deals with two architectures utilizing several physical properties provided by nanophotonics. First in Sect. 2, we discuss a memory-based architecture in which a large lookup table is recorded by configuring the size and positions of quantum dots, as well as individually implementing the required logical operation mechanisms for each table entry. Two basic functions are derived from this architecture. One is a data gathering, or summation, mechanism suitable for similarity evaluation, which is discussed in Sect. 3.

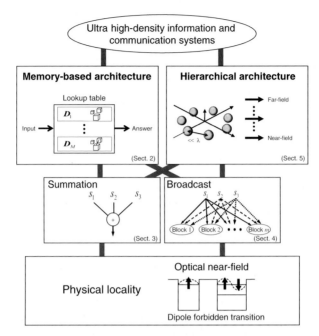

Fig. 1. Examples of system architectures utilizing the physics of nanophotonics, showing a memory-based architecture and a hierarchical architecture

As an extension of this summation architecture, Sect. 3 also discusses digital-to-analog conversion by configuring the coupling strength between QDs. The other basic function is broadcasting where query data is distributed to multiple table entries, as described in Sect. 4. We present enabling architectures by appropriate use of resonant energy levels between quantum dots and inter-dot interactions via optical near-fields that are forbidden for far-field light. Experimental results are also shown using CuCl quantum dots. In Sect. 5, we discuss hierarchical systems that use the different natures of light exhibited in optical near fields and far fields. The overall structure of this chapter is outlined in Fig. 1. Through such architectural and physical insights, we seek nanophotonic information and communications systems that can overcome the integration-density limit imposed by the diffraction limit of light with ultralow-power operation as well as unique functionalities which are only achievable using optical near-field interactions.

2 Nanophotonic Computing Architecture Based on High-Density Table Lookup

This section discusses the overall processing architecture based on high-density table lookup. We begin with a concrete example of a packet forwarding application, which is an important function in routers. In this application, the output port for an incoming packet is determined based on a routing table. For such functions, a content addressable memory (CAM) [6] or its equivalent is used; in a CAM, an input signal (content) serves as a query to a lookup table, and the output is the address of data matching the input. All optical means for implementing such functions have been proposed, for instance, by using planar lightwave circuits [7]. However, since we need separate diffraction-limited optical hardware for each table entry if based on today's known methods, if the number of entries in the routing table is on the order of 10,000 or more, the overall physical size of the system becomes unfeasibly large. On the other hand, by using diffraction-limit-free nanophotonic principles, huge lookup tables can be configured compactly.

First, we begin by relating the table-lookup problem to an inner-product operation. We assume an N-bit input signal $\boldsymbol{S} = (s_1, \ldots, s_N)$ and reference data $\boldsymbol{D} = (d_1, \ldots, d_N)$. Here, the inner product $\boldsymbol{S} \cdot \boldsymbol{D} = \sum_{i=1}^{N} s_i \cdot d_i$ will provide a maximum value when the input perfectly matches the reference data. However, the inner product alone is, in fact, not sufficient to determine correct matching of the input and reference. This can be demonstrated as follows. Assume, for example, a 4-bit input $\boldsymbol{S} = (1010)$ and two items of reference data $\boldsymbol{D}_1 = (1010)$ and $\boldsymbol{D}_2 = (1110)$. Both inner products result in a value of 2, but the correct matching data is only $\boldsymbol{D}_1$. Thus, the exclusiveness of the matching operations must also be considered. Correct matching can also be achieved by calculating the inner product of the *inverted* input signal and reference data. However, inversion is a difficult function to implement optically. One possible option is to properly design the modulation format [8], for instance, representing a logical level by two digits, such as Logic 1 = "10" and Logic 0 = "01". Then, an N-bit logical input is physically represented by $2N$ bits, which makes the inner product equivalent to the matching operation.

For packet data transfer, an operation known as longest prefix matching is important [9]. In this operation, a "don't-care" state is required. In the format described above, it can be simply coded by "11". Then, the resultant multiplication of a don't-care bit with an input bit will be 1 for either Logic 0 or 1.

Suppose that the reference data in the memory $\boldsymbol{D}_j$ $(j = 1, \ldots, M)$ and the input $\boldsymbol{S}$ are represented in the above format. Then, the function of the CAM will be to derive j that maximizes $\boldsymbol{S} \cdot \boldsymbol{D}_j$. A nanophotonic implementation of such a function can be implemented in a highly dense form, as shown in Sect. 3. In addition, a large array of such inner-product operations will allow a massively parallel processing system to be constructed.

Consequently, multiple inner products are equivalent to a matrix-vector multiplication, which is capable of implementing a wide range of parallel computations [4]. As a simple example, digital-to-analog conversion will be demonstrated by tuning the near-field interaction strength, as discussed in Sect. 3.2.

Furthermore, arbitrary combinational logic can be reformulated as a table-lookup operation; more specifically, any computation is equivalent to performing a lookup in a table where all possible input/answer combinations are

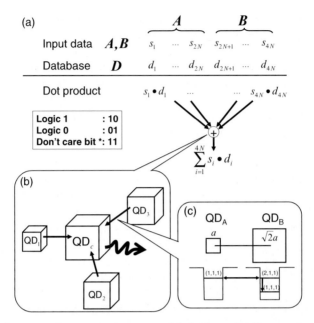

Fig. 2. (a) Inner-product operation as a table lookup. (b) Summation mechanism in quantum dots. (c) Inter-dot interaction via an optical near field

pre-recorded. For example, consider a two-input, two-bit ADD operation, $A + B$. In the ADD operation, the third-bit of the output (the carry bit) should be logical 1 when the second bits (that is, the 2^1 bit positions) of both inputs are 1, regardless of their first bits, that is, when $(A, B) = (1*, 1*)$ where $*$ denotes either 0 or 1 (i.e., a don't-care bit). Therefore, following the data representation format introduced above (Logic 1 = 10, Logic 0 = 01, and don't care = 11), the table-lookup entry D should be (10111011), so that any input combination satisfying $(A, B) = (1*, 1*)$ will provide a maximum inner product $S \cdot D$. This procedure is summarized in Fig. 2a.

3 Data Summation Using Near-Field Interactions

3.1 Data-Gathering Mechanism

As discussed in Sect. 2, the inner-product operations are the key functionality of the present architecture. The multiplication of two bits, namely $x_i = s_i \cdot d_i$, has already been demonstrated by a combination of three quantum dots [10, 11]. Therefore, one of the key operations remaining is the summation, or data-gathering scheme, denoted by $\sum x_i$, where all data bits should be taken into account.

In known optical methods, wave propagation in free-space or in waveguides, using focusing lenses or fiber couplers, for example, well matches such a data-gathering scheme because the physical nature of propagating light is inherently suitable for global functionality such as global summation. However, the level of integration of these methods is restricted due to the diffraction limit of light. In nanophotonics, on the other hand, the near-field interaction is inherently physically local, although functionally global behavior is required.

Here we implement a global data-gathering mechanism, or summation, based on the unidirectional energy flow via an optical near field, as schematically shown in Fig. 2b, where surrounding excitations are transferred towards a quantum dot QD_C located at the center [12, 13]. As a fundamental case, we assume two quantum dots QD_A and QD_B as shown in Fig. 2c. The ratio of the sizes of QD_A and QD_B is $1{:}\sqrt{2}$. There is a resonant quantized energy sublevel between those two dots, which are coupled by an optical near-field interaction [10, 11, 14], which allows the exciton population in the $(1, 1, 1)$-level in QD_A to be transferred to the $(2, 1, 1)$-level in QD_B [10, 14]. It should be noted that this interaction is forbidden for far-field light [15]. Since the intra-sublevel relaxation via exciton–phonon coupling is fast, the population is quickly transferred to the lower $(1, 1, 1)$-level in QD_B. Similar energy transfers may take place among the resonant energy levels in the dots surrounding QD_C so that energy flow can occur. One may worry that if the lower energy level of QD_B is occupied, another exciton cannot be transferred to that level due to the Pauli exclusion principle. Here, thanks again to the nature of the optical near-field interaction, the exciton population goes back-and-forth in the resonant energy level between QD_A and QD_B, a phenomenon which is known as optical nutation [10, 11, 14]. Finally, both excitons can be transferred to QD_B. The lowest energy level in each quantum dot is coupled to a free photon bath to sweep out the excitation radiatively. The output signal is proportional to the $(1, 1, 1)$-level population in QD_B.

Numerical calculations were performed based on quantum master equations in the density matrix formalism. The model Hamiltonian of the two dots is given by

$$H = \hbar \begin{pmatrix} \Omega_A & U \\ U & \Omega_B \end{pmatrix} , \qquad (1)$$

where $\hbar U$ is the optical near-field interaction, and $\hbar \Omega_A$ and $\hbar \Omega_B$, respectively, refer to the eigenenergies of QD_A and QD_B. For a two-exciton system, we can prepare seven states, as summarized in Fig. 3a, where one or two excitons occupy either one or two levels among the $(1, 1, 1)$-level in QD_A (denoted by A), the $(2, 1, 1)$-level in QD_B (denoted by B2), and the $(1, 1, 1)$-level in QD_B (denoted by B1). These seven states are interconnected either by inter-dot near-field coupling (U), exciton–phonon coupling (Γ), or relaxation to the radiation photon bath (γ_A for QD_A and γ_B for QD_B). Within the Born–Markov approximation of the Liouville equation [14, 16], we can derive multiple differential equations. In the following, we assume $U^{-1} = 50\,\text{ps}$, $\Gamma^{-1} = 10\,\text{ps}$, $\gamma_A^{-1} = 2\sqrt{2}\,\text{ns}$, and $\gamma_B^{-1} = 1\,\text{ns}$ as a typical parameter set.

First we consider an initial condition where there are two excitons in the system: one in QD_A and the other in QD_B (two-exciton system). The population of the $(1,1,1)$-level in QD_B corresponds to the output signal, which is composed of three states indicated by (i)–(iii) in Fig. 3a. The populations

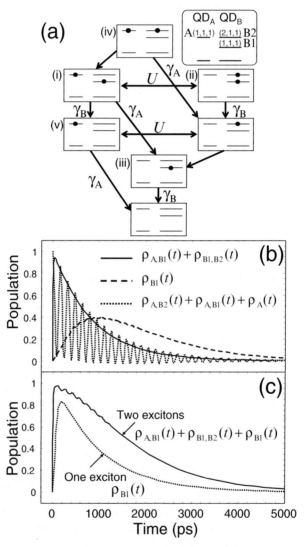

Fig. 3. (a) Bases of two-exciton system in two quantum dots coupled by optical near fields. (b) Time evolution of the population in a two-exciton system. (c) Population comparison between one- and two-exciton systems

for those three bases, which are diagonal elements of the density matrix, are, respectively, denoted by $\rho_{A,B1}(t)$, $\rho_{B1,B2}(t)$, and $\rho_{B1}(t)$, the first two of which, $\rho_{A,B1}(t)$ and $\rho_{B1,B2}(t)$, are related to the two-exciton dynamics of the system. They, respectively, show the time evolution of the one-exciton population in QD_A or in the upper level of QD_B, in addition to an exciton in the lower level of QD_B. The time evolution of $\rho_{A,B1}(t) + \rho_{B1,B2}(t)$ is shown by the solid curve in Fig. 3b. The other population, $\rho_{B1}(t)$, has just one exciton in B1, and so it represents the output evolution of the one-exciton system, which is shown by the dashed curve in Fig. 3b. Incidentally, the population when QD_A has an exciton, namely the sum of the populations related to bases (i), (iv), and (v) in Fig. 3a, is denoted by the dotted curve in Fig. 3b. Nutation is observed as expected since the lower level of QD_B is likely to be busy and the inter-dot near-field interaction is faster than the relaxation bath coupling at each dot.

We then compare the population dynamics between one- and two-exciton systems. The dotted curve in Fig. 3c shows the time evolution of the population in the lower level of QD_B, where, as initial conditions, one exciton exists only in QD_A. The solid curve in Fig. 3c is that for the two-exciton system. Physically the output signal is related to the integration of the population in the lower level of QD_B. Numerically integrating the population between 0 and 5 ns, we can obtain the ratio of the output signals between the two- and one-exciton systems as 1.86:1, which reflects the number of initial excitons. This is the summation mechanism.

A proof-of-principle experiment was performed to verify the nanoscale summation using CuCl quantum dots in an NaCl matrix, which has also been employed for demonstrating nanophotonic switches [10] and optical nanofountains [13]. We selected a quantum dot arrangement where small QDs (QD_1 to QD_3) surrounded a "large" QD at the center (QD_C), as schematically shown in Fig. 4a. Here, we irradiate at most three light beams with different wavelengths, 325, 376, and 381.3 nm, which, respectively, excite the quantum dots QD_1 to QD_3 having sizes of 1, 3.1, and 4.1 nm, respectively. The excited excitons are transferred to QD_C, and their radiation is observed by a near-field fiber probe. Notice the output signal intensity at a photon energy level of 3.225 eV in Fig. 4b, which corresponds to a wavelength of 384 nm or a QD_C size of 5.9 nm. The intensity varies approximately as 1:2:3 depending on the number of excited QDs in the vicinity, as observed in Fig. 4b. The spatial intensity distribution was measured by scanning the fiber probe, as shown in Fig. 4c, where the energy is converged at the center. Hence, this architecture works as a summation mechanism, counting the number of input channels, based on exciton energy transfer via optical near-field interactions.

Such a quantum-dot-based data-gathering mechanism is also extremely energy efficient compared to other optical methods such as focusing lenses or optical couplers. For example, the transmittance between two materials with refractive indexes n_1 and n_2 is given by $4n_1n_2/(n_1 + n_2)^2$; this gives a 4% loss if n_1 and n_2 are 1 and 1.5, respectively. The transmittance of an

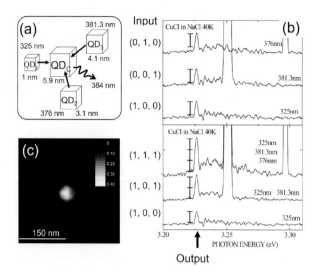

Fig. 4. Experimental results of the nanometric summation. (**a**) A quantum dot arrangement. (**b**) Luminescence intensity for three different numbers of excited QDs. (**c**) Spatial intensity distribution of the output photon energy

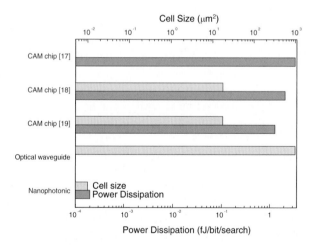

Fig. 5. Power dissipation and cell size comparison with electrical CAM VLSI chips and optical waveguides

N-channel guided wave coupler is $1/N$ from the input to the output if the coupling loss at each coupler is 3 dB. In nanophotonic summation, the loss is attributed to the dissipation between energy sublevels, which is significantly smaller. Incidentally, it is energy- and space-efficient compared to electrical CAM VLSI chips [17–19], as shown in Fig. 5.

We should also note, in terms of interconnections, that the input data should be commonly applied to all lookup table entries, which allows another possible interconnection mechanism. Since the internal functionality is based on energy transfer via optical near-field interactions and it is forbidden for far-field light, global input data irradiation, that is, broadcast interconnects, via far-field light may be possible; this is discussed in Sect. 4.

3.2 Digital-to-Analog Conversion Using Near-Field Interactions

In the summation mechanism shown in Sect. 3.1, the coupling strengths between the input QDs and the output QD are uniform. However, these coupling strengths can be independently configured, for instance, by modifying the relative distances. Theoretically, this corresponds to configuring U of the Hamiltonian in (1). For instance, consider three input QDs, QD_0 to QD_2, as schematically shown in Fig. 6a. By choosing U^{-1} of 410, 240, and 50 ps between QD_0 to QD_2 and the output QD, respectively, the simulated time evolution of the output population is as shown in Fig. 6b. The time integrals of the output populations originating from QD_0 to QD_2 between 0 and 5 ns are approximately in a ratio of 1:2:4. This leads to a digital-to-analog conversion formula given by

$$d = 2^0 s_0 + 2^1 s_1 + 2^2 s_2 , \qquad (2)$$

where d is the output, and s_0, s_1, and s_2 represent the presence/absence of excitations in QD_0 to QD_2, respectively. Here each of the inputs s_i is optically applied to the system, whose frequency is resonant with the $(1,1,1)$-level

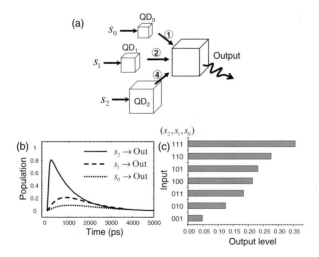

Fig. 6. (a) Digital-to-analog (DA) conversion. The near-field coupling is tuned so as to satisfy the relation for DA conversion. (b) Time evolution from each of the input bits to the output. (c) Experimental results of the output intensity level as a function of 3-bit input combinations

in QD_i. It should be noted that they are not coupled to the other QDs (i.e., input QD_j ($j \neq i$) and the output QD) since the corresponding energy levels are optically forbidden for the other QDs. Also, the initial state of the system is considered to be one in which an exciton is excited at each dot.

In the experiment, CuCl QDs in an NaCl matrix were used, as in Sect. 3.1, and three different input light frequencies were assigned to the three-bit input. Here, the output signal is considered to be the radiative relaxation from the lowest energy level of the output QD, which is observed with a near-field fiber probe at a wavelength of 384 nm. One remark here is that not every excited exciton produces the output signal; for instance there will be loss due to relaxation at each of the input QDs when the output energy level is occupied. However, such effects may not be serious since, as discussed in Sect. 3.1, nutation occurs among resonant energy levels and the relaxation rate at the output QD, which is the largest in the system in terms of size, is smaller than that at the input QDs. Figure 6c shows output signal intensity as a function of the presence (1) or absence (0) of the input excitation, as specified by (s_2, s_1, s_0), which were, respectively, 381.3, 376, and 325 nm. The output intensity is approximately linearly correlated to the input bit set combination, which indicates the validity of the digital-to-analog conversion mechanism. Compared to known optical approaches, such as those based on space–domain filtering and focusing lenses [3, 4], or optical waveguides and intensity filters [20], the nanophotonic approach achieves a significantly higher spatial density.

4 Broadcast Interconnects

Nanophotonics allows subwavelength scale device integration, but it imposes stringent interconnection requirements to couple external signals to nanophotonic devices. To fulfill such requirements, far- and near-field conversion has been pursued based on, for instance, plasmon waveguides [21, 22]. In this section, we show another interconnection scheme based on both far- and near-field interactions for data broadcasting purposes, as schematically shown in Fig. 7 [23].

As discussed in Sect. 2, data broadcasting is a fundamental operation found in memory-based architectures in which multiple functional blocks require the same input data, as schematically shown in Fig. 7a. Example of such architecture include matrix-vector product [3,4] and switching operations, such as in a broadcast-and-select architecture [24]. For example, consider a matrix-vector multiplication given by $v = As$, where $v = (v_1, \ldots, v_m)$ and $s = (s_1, \ldots, s_n)$, and A is an $m \times n$ matrix. Here, to compute every v_j from the input data s, broadcast interconnects are required if every v_j is calculated at distinct processing hardware. Optics is in fact well suited to such broadcast operations in the form of simple imaging optics [3, 4] or in optical waveguide couplers thanks to the nature of wave propagation. However, its integration density is physically limited by the diffraction limit, which leads to bulky system configurations.

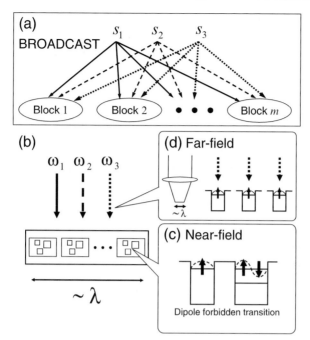

Fig. 7. (a) Broadcast-type interconnects and (b) their nanophotonic implementation. (c) Near-field interaction between quantum dots for internal functions. (d) Far-field excitation for identical data input (broadcast) to nanophotonic devices within a diffraction-limit-sized area

The overall physical operation principle of broadcasting is as follows. In nanophotonics, uni-directional energy transfer is possible between neighboring QDs via local optical near-field interactions and intrasublevel relaxation, as discussed in Sect. 3 and shown in Fig. 7c.

Suppose that arrays of nanophotonic circuit blocks, such as the nanophotonic switches described later, are distributed within an area whose size is comparable to the wavelength, as shown in Fig. 7b. Here, for broadcasting, multiple input QDs simultaneously accept identical input data carried by diffraction-limited far-field light by tuning their optical frequency so that the light is coupled to dipole-allowed energy sublevels, as illustrated in Fig. 7d and described in more detail later. In a frequency multiplexing sense, this interconnection method is similar to multiwavelength chip-scale interconnects [25]. Known methods, however, require a physical space comparable to the number of diffraction-limited input channels due to wavelength demultiplexing, whereas in our proposed scheme, the device arrays are integrated on the subwavelength scale, and multiple frequencies are multiplexed in the far-field light supplied to the device.

Here we explain the far- and near-field coupling mentioned earlier based on a model assuming CuCl QDs, which are employed in experiments described later. The potential barrier of CuCl QDs in an NaCl crystal can be regarded as infinitely high, and the energy eigenvalues for the quantized Z_3 exciton energy level (n_x, n_y, n_z) in a CuCl QD with side of length L are given by

$$E_{(n_x,n_y,n_z)} = E_{\mathrm{B}} + \frac{\hbar^2 \pi^2}{2M(L - a_{\mathrm{B}})^2} (n_x{}^2 + n_y{}^2 + n_z{}^2) , \qquad (3)$$

where E_{B} is the energy of the bulk Z_3 exciton, M is the mass of the exciton for the center-of-mass (CM) motion, a_{B} is its Bohr radius, n_x, n_y, and n_z are quantum numbers for the CM motion $(n_x, n_y, n_z = 1, 2, 3, \ldots)$, and $a = L - a_{\mathrm{B}}$ corresponds to an effective side length, taking into account the dead layer correction [26]. The exciton energy levels with even quantum numbers are dipole-forbidden states [15]. According to (3) there exists a resonance between the quantized exciton energy sublevel of quantum number $(1, 1, 1)$ for the QD with effective side length a and that of quantum number $(2, 1, 1)$ for the QD with effective side length $\sqrt{2}a$. For simplicity, we refer to the QDs with effective side lengths a and $\sqrt{2}a$ as "QD a" and "QD $\sqrt{2}a$", respectively. Energy transfer between QD, a, and QD, $\sqrt{2}a$, occurs via optical near fields, which is forbidden for far-field light [10, 11, 14].

We note that the input energy level for the QDs, that is, the $(1, 1, 1)$-level, can also couple to the far-field excitation. We utilized this fact for data broadcasting. One of the design restrictions is that energy-sublevels for input channels do not overlap with those for output channels. Also, if there are QDs internally used for near-field coupling, dipole-allowed energy sublevels for those QDs cannot be used for input channels since the inputs are provided by far-field light, which may lead to misbehavior of internal near-field interactions if resonant levels exist. Therefore, frequency partitioning among the input, internal, and output channels is important. The frequencies used for broadcasting, denoted by $\Omega_{\mathrm{i}} = \{\omega_{\mathrm{i},1}, \omega_{\mathrm{i},2}, \ldots, \omega_{\mathrm{i},A}\}$, should be distinct values and should not overlap with the output channel frequencies $\Omega_{\mathrm{o}} = \{\omega_{\mathrm{o},1}, \omega_{\mathrm{o},2}, \ldots, \omega_{\mathrm{o},B}\}$. A and B indicate the number of frequencies used for input and output channels, respectively. Also, there will be frequencies needed for internal device operations that are not used for either input or output (discussed later in the sum of product examples), denoted by $\Omega_{\mathrm{n}} = \{\omega_{\mathrm{n},1}, \omega_{\mathrm{n},2}, \cdots, \omega_{\mathrm{n},C}\}$, where C is the number of those frequencies. Therefore, the design criteria for global data broadcasting is to exclusively assign input, output, and internal frequencies, Ω_{i}, Ω_{o}, and Ω_{n}, respectively.

Figure 8 illustrates two examples of frequency partitioning, where the horizontal axis shows QD size and the vertical axis shows energy sublevels. The 3-digit sets in the diagram are the quantum numbers of the QDs. In an example shown in Fig. 8a, we used a nanophotonic switch (2-input AND gate) composed of three QDs with a size ratio of $1:\sqrt{2}:2$. The details of the switching principle are shown in [10]. The two input channels are assigned to QD a and QD $2a$, and the output appears from QD $\sqrt{2}a$. Here, multiple input dots QD a and QD $2a$ can accept identical input data via far-field light for broadcasting purposes. Adding more optical switches for different channels means adding different size dots, for instance, by multiplying the scale of the

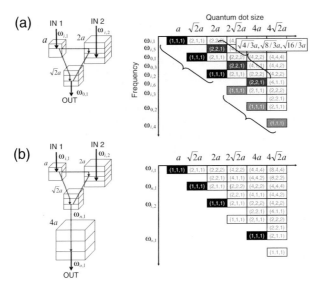

Fig. 8. Frequency partitioning among external and internal channels, and examples in (**a**) multiple implementations of 3-dot nanophotonic switches, and (**b**) 4-dot configuration for sum of products

QDs by a constant while maintaining the ratio $1:\sqrt{2}:2$, such as a QD trio of $2\sqrt{2}a$, $4a$, and $4\sqrt{2}a$, so that the corresponding far-field resonant frequencies do not overlap with the other channels. More dense integration is also possible by appropriately configuring the size of the QDs. As an example, consider a QD whose size is $\sqrt{4/3}a$. The $(1,1,1)$-level in this QD $\sqrt{4/3}a$ can couple to the far-field excitation. It should be noted that this particular energy level is equal to the $(2,2,1)$-level in QD $2a$, which is an already-used input QD; however, the far-field excitation in this particular energy level cannot couple to QD $2a$ since the $(2,2,1)$-level in QD $2a$ is a dipole-forbidden energy sublevel. Therefore, a QD trio composed of QDs of size $\sqrt{4/3}a$, $\sqrt{8/3}a$, and $\sqrt{16/3}a$ can make up another optical switch, while not interfering with other channels, even though all of the input light is irradiated in the same area.

Another situation where an internally used frequency exists is a sum of products operation. A simplified example is shown in Fig. 8b. The QD a and QD $2a$ operate on two inputs and their product appears in the $(1,1,1)$-level in QD $\sqrt{2}a$, which is further coupled to the sublevel $(4,2,2)$ in QD $4a$. The QD $4a$ is the output dot. Here, the QD $\sqrt{2}a$ is internally used; thus any frequency that could couple to QD $\sqrt{2}a$ cannot be used for other input channels.

To verify the broadcasting method, we performed the following experiments using CuCl QDs inhomogeneously distributed in an NaCl matrix at a temperature of 22 K. To operate a 3-dot nanophotonic switch (2-input AND gate) in the device, we irradiated at most two input light beams (IN1 and IN2). When both inputs exist, an output signal is obtained from the positions

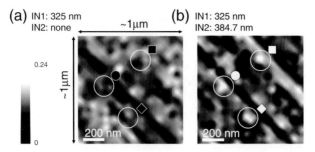

Fig. 9. Experimental results. Spatial intensity distribution of the output of 3-dot AND gates. (**a**) Output level: low (1 AND 0 = 0) and (**b**) output level: high (1 AND 1 = 1)

where the switches exist, as described above. In the experiment, IN1 and IN2 were assigned to 325 and 384.7 nm, respectively. They were irradiated over the entire sample (global irradiation) via far-field light. The spatial intensity distribution of the output, at 382.6 nm, was measured by scanning a near-field fiber probe within an area of approximately 1 μm × 1 μm. In Fig. 9a, only IN1 was applied to the sample where the output of the AND gate is ZERO (low-level), whereas in Fig. 9b both inputs were irradiated, which means the output is ONE (high-level). Note the regions marked by ■, ●, and ◆. In those regions, the output signal levels were, respectively, low and high in Fig. 9a and b, which indicates that multiple AND gates were integrated at densities beyond the scale of the globally irradiated input beam area. That is to say, broadcast interconnects to nanophotonic switch arrays are accomplished by diffraction-limited far-field light.

Combining this broadcasting mechanism with the summation mechanism discussed in Sect. 3 will allow the development of nanoscale integration of optical parallel processing devices, which have conventionally resulted in bulky systems.

5 Hierarchical Nanophotonic Systems

5.1 Hierarchy and Nanophotonics

One of the problems for ultra high-density nanophotonic systems is interconnection bottlenecks, which have been addressed earlier in Sect. 4 in terms of broadcast interconnects. Hierarchy is another perspective for solving interconnection issues in systems. One of the fundamental points differentiating nanophotnics from electronics is that nanophotonics is physically based on the flow of excitation, not electron transfer [27]. This can be demonstrated simply by the fact that a certain nanostructure can be observed differently depending on how we see it. For instance, an object that has a structure with subwavelength precision cannot be resolved by far-field light due to the diffraction limit of light, but can be resolved by a near-field optical microscope (NOM), as schematically shown in Fig. 10a and b. From a system perspective, therefore, we should deal with such hierarchical structures in nanophotonic systems in a systematic manner in future, but this section discusses a simple approach regarding a hierarchical optical memory system.

To achieve ultrahigh-capacity optical data storage, various methods to increase the storage density have been pursued, such as shortening the operating wavelength [28]. With such methods, the storage density is still bound by the diffraction limit of light. One technique to overcome this limitation is to use optical near fields [29]. These high-density optical memories, however, need certain seeking or scanning mechanisms, which might be a problem, for instance, when searching entire terabyte- or petabyte-scale memories. In dealing with this problem, we first note that information has hierarchy in terms of its meaning or quality, such as *abstract* and *detailed* information, *low* and *high* resolution information, and so forth. Similarly, as discussed below, we can find physical hierarchy in the different modes of light propagation. For example, in a near field, a spatial distribution of the individual dipole moments is obtained, whereas in a far field, the macroscopic features of the dipole moments are obtained. We associate these hierarchies in the system demonstrated in this section, that is, a hierarchical optical memory system having both near- and far-field readout functions with a simple digital coding scheme. As schematically shown in Fig. 10, in the far-field mode, low density, rough information is read-out, whereas in the near-field mode, high density, detailed information is read-out.

5.2 Physical and Logical Hierarchy Using Nanophotonics

The two-layer hierarchical memory in this section is explained using the notations *far-code* and *near-code*. The *far-code* depends on the array of bits distributed within a certain area and is determined logically to be either ZERO or ONE. Each *far-code* is comprised of multiple smaller-scale elements, whose existence is determined by the *near-code*. To obtain such information hierarchically, we introduce the following simple logical model.

Consider an $(N + 1)$-bit digital code, where N is an even number. Now, let the far-code be defined depending on the number of ONEs (or ZEROs) contained in the $(N + 1)$-bit digital code:

$$\text{far} - \text{code} = \begin{cases} 1 \text{ if the number of ONEs} \geq N/2 \,, \\ 0 \text{ otherwise} \,. \end{cases} \tag{4}$$

The $(N + 1)$ digits provide a total of 2^{N+1} possible different permutations, or codes. Here, we note that half of them, namely 2^N permutations, have less than $N/2 + 1$ ONEs among the $(N + 1)$ digits(i.e., far-code $= 1$), and the other half, also 2^N permutations, have more than $N/2 + 1$ ZEROs

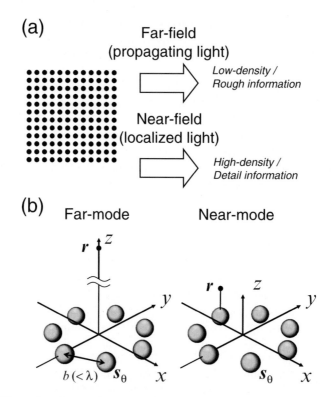

Fig. 10. Hierarchy of nanophotonics. Two layer hierarchical optical memory system in which near fields and far fields read detailed dipole distributions and features within a region-of-interest, respectively

(a)

Given 8-bit code	*Near-code* (*far-code* = 0)	*Near-code* (*far-code* = 1)
0	000000000 (0)	000011111(31)
1	000000001 (1)	000101111(47)
2	000000010 (2)	000110111(55)
⋮	⋮	⋮
92	001101010 (106)	110001011(395)
93	001101100 (108)	110001101(397)
94	001110000 (112)	110001110(398)
⋮	⋮	⋮
253	111001000 (456)	111111101(509)
254	111010000 (464)	111111110(510)
255	111100000 (480)	111111111(511)

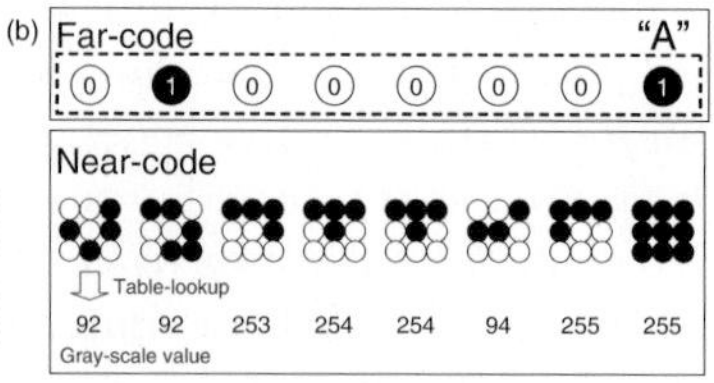

Fig. 11. Example of logical model for the near-code and far-code. Here, the original 8-bit information is coded differently in the near-code depending on its corresponding far-code, which is either ZERO or ONE

(i.e., far-code = 0). In other words, 2^N different codes could be assigned to two $(N + 1)$-bit digital sequences so that their corresponding far-codes are ZERO and ONE, respectively. We call this $(N + 1)$-bit code a near-code.

In Fig. 11, example near-codes are listed when $N = 8$. The correspondence between 2^N original codes and the $(N + 1)$-bit near-codes is arbitrary. Therefore, we need a table-lookup when decoding an $(N + 1)$-bit near-code to the original code. The example near-codes shown in Fig. 11a are listed in ascending order, but other lookup-tables or mappings are also possible.

Figure 11b schematically demonstrates example codes in which a 9-bit near-code is represented in a 3×3 array of circles, where black and white mean ONE and ZERO, respectively. Here, (1) if the number of ONEs in the near-code is larger than five, then the far-code is ONE; and (2) if the number of ONEs in the near-code is four or less, then the far-code is ZERO.

Suppose, for example, that the far-code stores text data and the near-code stores 256-level (8-bit) image data. Consider a situation where the far-code should represent an ASCII code for "A," whose binary sequence is "0100001." Here, we assume that the gray levels of the first two pixels, which will be coded in the near-code, are the same value. Here, they are at a level of "92." However, the first two far-codes are different (ZERO followed by ONE). Referring to the rule shown in Fig. 11a, and noticing that the first far-code is ZERO and the near-code should represent "92," the first near-code should be "001101010." In the same way, the second near-code is "110001011," so that it represents the level "92," while its corresponding far-code is ONE.

The far-code is determined based on the rule given by (4), which depends on the number of ONEs coded in the near-code. Here, we employ a simple physical model where the near-code is represented by an array of dipole moments. As schematically shown in Fig. 10b, dipole moments are distributed in an xy plane, where an $(N + 1)$-bit code is assigned in an equally spaced grid. In Fig. 10b, each of the dipole moments is assumed to be induced in a small particle which is placed on a circle, and the distance between adjacent particles is b. The electrical field at position r in Fig. 10b is given by

$$E(\boldsymbol{r}) = \sum_{\theta} E_{\theta} e^{-i\omega t + ik|\boldsymbol{r} - \boldsymbol{s}_{\theta}|} \frac{1}{|\boldsymbol{r} - \boldsymbol{s}_{\theta}|} , \qquad (5)$$

where ω is the operating frequency, k is the wave number, and $\boldsymbol{s}_{\theta}$ represents the position of a dipole specified by index θ [29]. The existence of the dipole at the position $\boldsymbol{s}_{\theta}$ is given by the near-code as

$$E_{\theta} = \begin{cases} 0 & \text{near-code}(\theta) = 0 , \\ E_0 & \text{near-code}(\theta) = 1 . \end{cases} \qquad (6)$$

Suppose that the pitch of adjacent dipoles is b. Here, if we assume that $b \ll 1 \ll r$, then (5) is simplified to

$$E(\boldsymbol{r}) = E_0 \frac{e^{-i\omega t + ikr}}{r} \sum_{\theta} \text{near-code}(\theta) , \qquad (7)$$

which means that the electrical field intensity at position r is proportional to the number of ONEs given by the near-code in that area.

Simulations were performed assuming ideal isotropic metal particles to see how the scattering light varies depending on the number of particles for the far-code using a finite difference time domain-based simulator (*Poynting for*

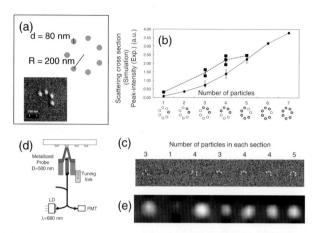

Fig. 12. (**a**) Each section consists of small particles. (**b**, *circular marks*) Scattering cross-sections are calculated depending on the number of particles in each section. (**c**) SEM picture of an Au particle array. (**d**) Experimental setup. (**e**) Intensity profile captured for the *far-code*. (**b**, *square marks*) Peak-intensity of each section

Optics, a product of Fujitsu, Japan). Here we assume that 80-nm-diameter particles are distributed over a 200-nm-radius circular grid with a constant interval, as shown in Fig. 12a. The solid circles in Fig. 12b show calculated scattering cross-sections as a function of the number of particles. The assignment of particle(s) in the grid is also shown. A linear correspondence to the number of particles was observed. This result supports the simple physical model described above.

In order to experimentally demonstrate such principles, we obtained far-field intensity distributions of an array of Au particles, each with a diameter around 80 nm, was distributed over an SiO_2 substrate in a 200-nm-radius circle. These particles were fabricated by a liftoff technique using electron-beam (EB) lithography with a Cr buffer layer. Each group of Au particles was spaced by 2 μm. An SEM image is shown in Fig. 12c in which the values indicate the number of particles within each group. In order to illuminate all Au particles in each group and collect the scattered light from them, we used a near-field optical microscope (NOM) with a large-diameter-aperture (500 nm) metallized fiber probe, as shown in Fig. 12d, in an illumination collection setup. The light source used was a laser diode with an operating wavelength of 680 nm. The distance between the substrate and the probe was maintained at 750 nm. Figure 12e is an intensity profile captured by the probe, by which the information in the far-mode is retrieved. The solid squares in Fig. 12d indicate the peak intensity of each section, which increased linearly. Since the signal level difference in the far-mode is small when the difference in the number of particles is small, a better way of resolving signals in the far-mode will be required. However, the experimental results shown here represent one fundamental architecture that we can exploit the physical hierarchy achievable by nanophotonics.

To summarize this section, a hierarchical optical memory system is presented in which near-fields are used to read detailed dipole distributions, whereas far fields are used to detect features within a region-of-interest. Simulations and experimental results were also shown. With hierarchical coding, near- and far-field accesses are associated with different hierarchical information which should help to overcome problems involved in searching huge memory spaces, and other applications. General analysis and design issues for hierarchical nanophotonic systems are presently under investigation.

6 Summary and Discussion

Two architectural approaches to nanophotonic information and communication systems are discussed in this chapter. One is a memory-based architecture which is based on table lookup using near-field interactions between QDs. In addition, content addressable memories, digital logic, and matrix-vector multiplication can be implemented in this architecture. As fundamental functional elements, a data summation mechanism, digital-to-analog conversion, and broadcast interconnects are presented, and their proof-of-principle experiments are demonstrated using CuCl QDs. Owing to its high spatial density and low power dissipation, a massive array of such functional components will be useful in applications such as massive table-lookup operations in networking and information processing systems. Another approach is one focusing on hierarchy. As an example, a hierarchical memory system is presented with simulations and experimental results using Au particles distributed on a subwavelength scale.

Generality of processing remains an open issue since it requires random access memories; other design strategies, such as binary decision diagrams [30], are other possible candidates. Delay-line buffers are also extremely important for optical networks to replace the bulky optical fiber loops used for buffering [31]; the possibility of using nanophotonic devices in these applications is now being pursued [11]. The extremely high spatial density will also lead to novel system design concepts, for instance, in redundancy or fault tolerance [32]. In addition, further investigation of system application issues is indispensable, such as interconnections, hierarchical architectures, the fabrication limitations of nanostructures [33], and new applications unachievable by other technologies.

References

1. For example, *2004 WHITE PAPER, Information and Communications in Japan.* (Ministry of Internal Affairs and Communications (MIC), Japan)
2. S. Yao, B. Mukherjee, S. Dixit: IEEE Commun. Mag. **38**, 84 (2000)
3. J.W. Goodman, A.R. Dias, L.M. Woody: Opt Lett. **2**, 1 (1978)
4. P.S. Guilfoyle, D.S. McCallum: Opt. Eng. **35**, 436 (1996)

5. M. Ohtsu, K. Kobayashi, T. Kawazoe, S. Sangu, T. Yatsui: IEEE J. Select. Top. Quantum Electron. **8**, 839 (2002)
6. H. Liu: IEEE Micro **22**, 58 (2002)
7. A. Grunnet-Jepsen, A.E. Johnson, E.S. Maniloff, T.W. Mossberg, M.J. Munroe, J.N. Sweetser: Electron. Lett. **35**, 1096 (1999)
8. M. Naruse, H. Mitsu, M. Furuki, I. Izumi, Y. Sato, S. Tatsuura, M. Tian, F. Kubota: Opt. Lett. **29**, 608 (2004)
9. K. Kitayama, M. Murata: IEEE J. Lightwave Technol. **21**, 2753 (2003)
10. T. Kawazoe, K. Kobayashi, S. Sangu, M. Ohtsu: Appl. Phys. Lett. **82**, 2957 (2003)
11. S. Sangu, K. Kobayashi, A. Shojiguchi, M. Ohtsu: Phys. Rev. B **69**, 115334 (2004)
12. M. Naruse, T. Miyazaki, F. Kubota, T. Kawazoe, K. Kobayashi, S. Sangu, M. Ohtsu: Opt. Lett. **30**, 201 (2005)
13. T. Kawazoe, K. Kobayashi, M. Ohtsu: Appl. Phys. Lett. **86**, 103102 (2005)
14. T. Kawazoe, K. Kobayashi, J. Lim, Y. Narita, M. Ohtsu: Phys. Rev. Lett. **88**, 067404 (2002)
15. Z.K. Tang, A. Yanase, T. Yasui, Y. Segawa, K. Cho: Phys. Rev. Lett. **71**, 1431 (1993)
16. H.J. Carmichael: *Statistical Methods in Quantum Optics I.* (Springer, Berlin Heidelberg New York 1999)
17. I. Arsovski, A. Sheikholeslami: A current-saving match-line sensing scheme for content-addressable memories. In: *2003 IEEE International Solid-State Circuits Conference, Digest of Technical Papers.* Vol. 1, pp. 304–494
18. I. Arsovski, T. Chandler, A. Sheikholeslami: IEEE J. Solid-State Circuits **38**, 155 (2003)
19. P.-F. Lin, J.B. Kuo: IEEE J. Solid-State Circuits **36**, 666 (2001)
20. T. Saida, K. Okamamoto, K. Uchiyama, K. Takiguchi, T. Shibata, A. Sugita: Electron. Lett. **37**, 1237 (2001)
21. T. Yatsui, M. Kourogi, M. Ohtsu: Appl. Phys. Lett. **79**, 4583 (2001)
22. J. Takahara, Y. Suguru, T. Hiroaki, A. Morimoto, T. Kobayashi: Opt. Lett. **22**, 475 (1997)
23. M. Naruse, F. Kubota, T. Kawazoe, S. Sangu, K. Kobayashi, M. Ohtsu: Optical interconnects using optical far- and near-field interactions for high-density data broadcasting. In: *Conference on Lasers and Electro-Optics (CLEO) 2005.* Paper CWF6
24. B. Li, Y. Qin, X. Cao, K.M. Sivalingam: Opt. Networks Mag. **2**, 27 (2001)
25. E.A. De Souza, M.C. Nuss, W.H. Knox, D.A.B. Miller: Opt. Lett. **20**, 1166 (1995)
26. N. Sakakura, Y. Masumoto: Phys. Rev. B **56**, 4051 (1997)
27. M. Ohtsu, H. Hori: *Near-Field Nano-Optics* (Kluwer Academic/Plenum Publishers, New York 1999)
28. For example, http://www.blu-ray.com/
29. M. Ohtsu, K. Kobayashi: *Optical Near Fields.* (Springer, Berlin Heidelberg New York 2004)
30. S.B. Akers: IEEE Trans. Comput. **C-27**, 509 (1978)
31. D.K. Hunter, M.C. Chia, I. Andonovic: IEEE J. Lightwave Technol. **16**, 2081 (1998)
32. K. Nikolic, A. Sadek, M. Forshaw: Nanotechnology **13**, 357 (2002)
33. T. Yatsui, S. Takubo, J. Lim, W. Nomura, M. Kourogi, M. Ohtsu: Appl. Phys. Lett. **83**, 1716 (2003)

Index

Springer Series in
OPTICAL SCIENCES

Springer Series in
OPTICAL SCIENCES